Barns of Tennessee

The Tennessee Electric Cooperative Association and its official publication, *The Tennessee Magazine*, proudly present *Barns of Tennessee*. This project began in 2007 with a request to readers of *The Tennessee Magazine* to share photographs and stories of their beloved barns.

Acres of farmland and their once essential barns continue to disappear from the landscape as time and development overtake them. With this publication, it's our hope to document the histories of various types and styles of Tennessee barns, possibly bring back some memories of a simpler rural lifestyle, and help preserve some barns not yet lost.

It is traditional that farmers and small-town residents have always found it advantageous to work together for their mutual and common good. In that spirit, rural electric cooperatives began in Tennessee some 70 years ago.

In 1935, approximately 90 percent of people living in town had electricity while only 4 percent of Tennessee's rural population did. At that time, most people living on farms or in rural areas and small towns could not get electric service from existing electrical distributors because it was not profitable enough to build the lines.

Franklin Delano Roosevelt signed the Rural Electrification Act in 1935 to help bring electricity to rural America. As a result, rural electric cooperatives, owned by the members with a mission to serve the members, formed across the nation. Affordable, reliable electricity was finally a reality for rural America. It revolutionized farmlife.

Today, every American is touched in some way by cooperative enterprises. Approximately 40,000 separate cooperative organizations, with more than 50 million members, currently operate in this country. Statistics show that, of these 40,000 cooperatives, approximately 6,700 are member-owned, such as our electric cooperatives.

Tennessee's 23 rural electric systems are in step with the balanced progress that is making a great state an even greater one. Rural electric systems came into being and have progressed to their present state because they were designed for service to many people rather than financial profit for a few.

Second Printing 2009

For information, please write:
The Donning Company Publishers
184 Business Park Drive, Suite 206
Virginia Beach, VA 23462

TECA
General Manager
Tom Purkey

Project Director
Neil Hendricks

Editors
Robin Conover
Chris Kirk
Jerry Kirk
Trish Milburn

Designer
Ron Bell

Library of Congress
Cataloging-in-Publication Data
Hankins, Caneta Skelley.
Barns of Tennessee / Caneta Skelley Hankins
and Michael Thomas Gavin.
p. cm.
"With photographs from readers of the Tennessee
Magazine, a publication of the Tennessee Electric
Cooperative Association."
Includes bibliographical references and index.
ISBN 978-1-57864-545-9 (hardcover : alk. paper)
1. Barns--Tennessee--Pictorial works.
I. Gavin, Michael Thomas.
II. Tennessee magazine. III. Title.
NA8230.H34 2009
728'.92209768--dc22
2008053917

Printed in the United States of America

Weathered by time, this barn no longer stands. It was located along Highway 68 in East Tennessee.
Photograph by Robin Conover.

Contents

About the covers: Information about the photographs included on the front, back and inside covers can be found on pages 150-151.

STAY
ON
U.S.
64
and
See

Barns of My Youth

I miss the barns of my youth,
The ones that read, "See Rock City."
Hungry Caterpillars ate them alive,
Spitting out nails
To become thorns in the side
Of crawling asphalt,
Erasing small towns
To link big cities.
They die hard, these old barns,
Leaning on the everlasting
Shoulders of Time
That cushion the fall of rotting boards.
Light seeps through decaying skeletons,
Causing shadows to tiptoe
Like ballerinas dancing the waltz of the wind.
In our hurry to get there
We destroyed the landscape:
Masterpieces of America.

– Margaret Britton Vaughn
Poet Laureate of Tennessee

Preface and Acknowledgments

From the beginning, we were challenged and intrigued by this special project that pays homage to the character, architecture, and history of the barns of Tennessee. Although change is inevitable, many of us are unwilling to let the things that were and still are important disappear from our cultural memory and, quite literally, from our sight. Barns, the largest of farm buildings and the most representative architectural statement in an agricultural environment, lead the way in endangered resources. For nearly 200 years, agriculture was the number one industry, and its impact and influence in every aspect of the state's history can hardly be overemphasized. As Tennessee moves into its third century, however, sprawl and suburbanization are consuming huge amounts of previously rural acreage at an ever-increasing pace. Large farms are broken into tracts, and the houses, barns, and outbuildings that stood for years either quickly disappear or are allowed to slowly decay. Families that farmed for decades can no longer make a living from farming and, for many reasons, decide to sell the homestead.

On the other hand, many farmers are determined to retain and preserve their land, their homes, and their barns. These men and women are the keepers of Tennessee's agrarian past. They carry on agricultural traditions; they diversify their crops and livestock; they continuously study and learn to succeed in the business of farming; and they conserve buildings, the soil, green spaces, wooded areas, and waterways. For their decisions and their efforts, these barns and their owners deserve to be recognized and appreciated.

The Tennessee Electric Cooperative Association (TECA) is the sponsor of this publication, and a request for photographs appeared in the summer 2007 issues of its popular *Tennessee Magazine*. Soon images of barns were delivered daily to the Nashville headquarters in envelopes, packages, and boxes. The process of sorting and organizing more than 3,000 photographs was daunting, but viewing and reading each submission was a privilege and a pleasure. As we marveled at the response, one thing became clear: Tennesseans love their barns!

Selecting only 375 photographs was agonizing. Each submission had merit, and many people sent in stories, poems, essays, and reminiscences to further emphasize the importance of the subject. Choices, however, had to be made. Because this is a book about Tennessee barns, all 95 counties are represented. Hundreds of photographs were submitted for some counties, just a few from others, and none from five. In the latter situation, or when a specific illustration was needed to interpret a topic, we searched for appropriate photographs in existing collections, usually that of the Center for Historic Preservation at Middle Tennessee State University (MTSU), which includes the Tennessee Century Farms files.

In most cases, when a submission included specific information about the barn such as its age, builders, uses, and owners, we chose that photograph over those that provided no details. When there were multiple submissions of the same structure, we selected the photograph that provided the best perspective, contrast, detail, or resolution. Though some individuals submitted an impressive array of pictures, the format was limited to three entries from the same person. For each image chosen, the submitter and the photographer are credited. When the names of barn owners were provided on the entry form, we usually included that information along with the county in which the barn is located. Given the production schedule of this publication, we could not do additional research on all the entries, but we have made every effort to provide accurate information and to spell names and places correctly, based on the details provided with the photograph. When quotations are designated in the captions, they have been taken directly from the entry forms.

As the authors, it was our role and professional responsibility to provide a substantive architectural, agricultural, and historical context for this book rather than just organize a visually pleasing publication. Each entry was studied carefully several times before we finally selected those photographs that would best illustrate and interpret the complex story of Tennessee's barns. The bibliography includes books and articles as well as Web sites that were consulted in the preparation of the manuscript.

The Web site of the United States Department of Agriculture, National Agricultural Statistics Service, Tennessee Field Office, was an especially helpful resource for current facts and figures. Our thanks go to Dr. George

F. Grandle, interim head of the Biosystems Engineering and Soil Service Department at the University of Tennessee, who kindly provided us with quality copies of several agricultural building plans and offered valuable insights on the UT Extension chapter.

To those individuals whose submissions are included in this book and to those whose entries are not a part of this publication, we are sincerely grateful for your interest in this project and for your effort and expense. As a group, you have provided an invaluable collection of photographs that is an excellent, though still incomplete, survey of Tennessee barns in the early 21st century. More great barns exist, and they deserve to be photographed and recorded in their own counties. With the changes that are taking place and those that will continue to transform the state in the future, this documentation of our evolving lifestyle, economy, and land use will grow more significant with each decade. This collection will be used by scholars, students, and researchers for years to come. Thank you for contributing to this photographic record of barns, which will be archived by TECA for the foreseeable future.

To Tom Purkey, general manager of the Tennessee Electric Cooperative Association, and the staff at TECA headquarters, we thank you for your kind hospitality and the generous work space you provided. We especially appreciate the working partnership and assistance of Robin Conover, professional photographer and editor of *The Tennessee Magazine*, throughout this entire process.

Dr. Carroll Van West, director of the MTSU Center for Historic Preservation, supported this project from its inception. We are most grateful for his patience and encouragement, and also for providing photographs from his own collection. To our colleagues at the Center, we appreciate your tolerance and enthusiasm.

For his thoughtful comments, based on his experience and knowledge of Tennessee agriculture, we are indebted to James Nance, who graciously reviewed this manuscript.

Finally, with sincere respect and gratitude, we salute the readers of *The Tennessee Magazine* who, along with several Tennessee Century Farm families, generously shared their photographs, memories, and stories of barns. We hope you believe this product is worthy of its subject and that, in the best tradition of so many of Tennessee's barns, it will stand the test of time.

– *Caneta Skelley Hankins*

– *Michael Thomas Gavin*

The glass was long ago broken out of this window as time and vandals took their toll on this Hereford farm dairy barn in Kelso. *Photograph by Robin Conover.*

Introduction

Barns — whether large, medium, or small; red, white, or green; new, old, or in-between — endure as the most familiar symbol of an agricultural lifestyle. Barns can tell stories, too, if we take the time to observe their age, design, materials, method of construction, and if we listen to those who know their history. Questions about where different kinds of farm structures first appeared, how they came about, what they were used for, who used them, and why they were either kept or abandoned can be answered sometimes by examining the faint evidence that lingers both outside and within the buildings. These surviving physical traces can be augmented by the traditional tales, historic images, personal accounts, memorabilia, and other documents handed down through the years within individual families and the collective rural community.

In Part I, barns are grouped into categories based on building methods and architectural types in order to compare and contrast their similarities and differences as well as to discuss briefly their appearance and function, though those aspects are more fully treated in subsequent chapters. Part II explores the relationship of barns to their traditional agricultural uses as shelters for animals, storage for crops, and work spaces for people engaged in various farming tasks. Part III invites you to enjoy barns that are used as advertisements and displays for objects, signs, art, and flags. The sections within this part also provide comparisons between the appearance of barns in past and present incarnations, describe some new and different uses, and mention a few of the options that are available to owners to help them hold on to their barns. Finally, we take a look at barns whose futures are uncertain and end our tribute to barns with images of a few of those that no longer exist.

Until lately, America's barns have been easily accessible artifacts that directly linked the past and the present. The situation has changed so drastically in such a short period that Wallace Yoder, president of the Illinois-based BarnKeepers, which was formed to promote the documentation, restoration, and preservation of barns, recently told the Associated Press that "Someday, we'll wake up and the barns are all gone, and it's gonna be tough to educate our future generations."

This statement also applies to Tennessee's agricultural environment today. Many of the state's residents are now two, three, or more generations removed from the farm and have no first-hand knowledge of the complexity and the necessity of agriculture. Further, a large percentage of the population does not fully appreciate the economic contributions of agriculture to our state ($44.2 billion and 342,000 jobs in 2007). Increasingly, we depend on local farmers for an available, affordable, and safe food supply. With that in mind, this book is an effort to educate, to remember, and to consider thoughtful and sensible measures for the future. Barns have much to tell about the lifestyle, culture, economy, technology, needs, and concerns of past and current generations, and their conservation and continued use will have an undeniable effect on the future. Long may they stand.

Right, a large, multipurpose barn near Rogersville in Hawkins County.
Photographed by Robin Conover.

Part I

Barn Architecture and History

Introduction

Everybody knows what a barn looks like. In reality, though, no two barns are exactly the same. Some of these differences are relatively trivial such as what color the structure is painted. But other distinctions are more profound and deal with the origins, distribution, and purposes of the farm buildings. Asking and answering questions like these provide us with an appreciation of a rural lifestyle that is rapidly fading. Tennessee is a state with a solid agricultural base that historically has provided both sustenance and satisfaction for most of its citizens. Any building that a farmer worked in and used for specific tasks had good reasons for its appearance and internal structure. The goal of a student of rural architecture is to distinguish the important characteristics of these buildings among the many that can be observed and use this knowledge to establish different barn types based on their function and form. Part I is an attempt to differentiate among the complex variety of both barn building methods and barn types. The materials used for construction — such as logs, rock, and others — do not define building types; they represent methods of building. Usually, building types can be composed of any kind of material, sometimes in combination with others. Hopefully, the following narrative can lead to a better understanding of the incredible diversity of Tennessee's historic agricultural buildings.

This wooden door, with a hand-carved wooden hinge, hangs on a corncrib in the Cable Mill Historic Area of Cades Cove in the Great Smoky Mountains National Park. *Photographed by Robin Conover.*

Log Barns

Left, a V-notched cedar log drive-in crib barn located near Martin Springs in Marion County was expanded by the addition of a large frame hayloft and an open side shed.
Submitted and photographed by Rebecca Sliger.

Below, constructed in the 19th century with chestnut logs, Brett and Liz Beavers' sturdy four-crib barn rests along the side of Mill Creek in Hickman County.
Submitted and photographed by Brett Beavers.

The early Anglo-American settlers ventured to Tennessee for a variety of reasons, but all sought fresh opportunities in the land over the mountains. This vast forested area became a meeting ground for adventurous farmers and their families from Pennsylvania to the Carolinas. Various cultural traditions (British, Germanic, and Scandinavian, among others) came together on the frontier. Constant experimentation to find solutions for new challenges led to the development of larger, complex, and more useful buildings.

Food and shelter were the main priorities of the pioneers, and they began work as soon as they arrived at their destination. Among their first tasks were to clear a few acres and plant a corn crop among the deadened trees and stumps. They allowed domestic animals such as swine and cattle to forage for themselves in the nearby canebrakes. Corn was the most important crop and served as the basis of subsistence agriculture during much of the 19th century. The storage and protection of this basic commodity were essential for survival.

Virgin timber that needed to be cut down for fields and pastures provided the raw material to build the first buildings on the new farmstead. In the absence of sawmills to cut the trees into usable lumber, farmers turned to log construction as a practical alternative. A rectangular structure made of logs (known as either a pen or a crib) was extremely adaptable, and similar units could be added to the first in any number of ways. With multiple buildings housing different agricultural activities, many Southern farms took on the appearance of a little village.

Early barns were modest structures, built out of horizontal logs and notched at the corners. Some carpenters hewed the timbers flat on two sides, but others left them in the round. The exterior walls of crib barns (which commonly ranged in length from 6 to 20 feet) were left unsided, and the cracks between the logs seldom were chinked. Tall side walls permitted extra space for hay storage above.

The simplest farm building consisted of a single log crib covered by a steeply pitched gable roof of either thatch or long wooden shingles (called boards by the builders). The first settlers sometimes attached a log crib to one end of their own houses in order to keep a close eye on the family's milk cow, but after the Native American threat had ended, this forced arrangement quickly faded away.

The single-pen log crib, with either a shed or double-pitched roof and a door in the gable end, primarily served as a storage place for corn. The narrow spaces between the stout logs provided both ventilation and a measure of security. The small size of these barns made it difficult to do much farm work inside. The simplest method of expansion was to attach a lightweight frame shed with a low-pitched roof to either one or more sides of the log crib. Two important variations of the single-crib barn are the side-drive crib and the front-drive crib. Both types have an open runway adjoining one of the sides of the building to permit the unloading of corn out of the weather. When not in use, the sheltered area could be used to protect equipment, while the loft provided space for hay storage. Later, some farmers even enclosed the runway.

Crib barns were easily expandable. When the owner needed more space, it was relatively simple to build another crib a short distance from the first and cover the whole structure with a common roof. This approach resulted in a whole new family of larger barns known as double-cribs. The different types were distinguished from each other primarily by the configuration of their roof systems. The three main varieties of the double-crib barn types are eaves-front, gable-front, and cantilevered (the cantilevered barns are discussed in another chapter).

The hardwood logs composing the side walls of this Hickman County barn located on Swan Creek Road stretch a full 42 feet long. Two log walls divide the interior into three sections. The raised roof with the hay hood is not original. This barn no longer stands. *Courtesy of the MTSU Center for Historic Preservation.*

Divided into three parts with the ridgeline perpendicular to the runway, the eaves-front double-crib barn resembles the classic English barn on the exterior. This arrangement permits additional units to be placed on either side of the original building. Sometimes farmers built three, four, or more cribs in a line under the same roof divided by their runways. This linear configuration was particularly well-suited for stables.

On the other hand, the gable-front double-crib's ridgeline is parallel to the runway. Additional cribs could be attached at the rear of the barn, increasing its depth. When a second double-crib was placed facing the first one and separated by another runway, the result was a four-crib barn with the two runways forming a cross-like shape. This novel barn type originated in the eastern part of the state during the early settlement period and spread westward from there. Four-crib barns are particularly common in the Tennessee River Valley region to this day.

A typical fire-cured tobacco barn can be visited at the USDA Forest Service's "The Homeplace" in the Land Between The Lakes National Recreation Area in Stewart County. Note the owl holes cut into the gable siding. *Submitted and photographed by Nancy Adgent.*

Burley tobacco hangs from the tier rails that are visible at the gable end of the Jackson County barn that John Cox built back in 1890. Note the large opening for ventilation above the top course of the V-notched logs. *Submitted by Ashley Crabtree. Photographed by Josh Crabtree.*

However, many found this runway plan to be inefficient because it wasted so much space. Savvy builders dealt with this shortcoming by enclosing both open areas along the side walls. This alteration left a long central aisle parallel to the ridgeline flanked by log cribs on both sides. The improved version of the four-crib barn is called a central aisle barn (or a transverse-crib barn). Its suitability and ease of construction quickly became recognized throughout Tennessee and eventually well beyond its borders (please refer to the Central Aisle Barns chapter for additional information).

Some log barns could be quite big. Huge trees, 40 feet and taller, were cut down and hewn for the side walls of these barns. Often the large crib was divided into two or more sections by internal log walls. These massive farm buildings are some of the largest log structures ever built in the state.

A description of historic log farm buildings would not be complete without the mention of a particular kind of building that serves a special purpose: the tobacco barn. Tobacco barns provide a protected space in which to dry the crop. Two main methods of processing ripe tobacco — dark-firing and air-curing — are employed in the Volunteer State, and each requires a barn with specific characteristics. A more detailed discussion of tobacco culture can be found in Part II of this publication.

Just before the Civil War, J. G. W. Jimerson erected a fine drive-in crib barn in Obion County, just northwest of Troy. The unique entrance porch shown here rarely is found on log farm buildings. *Submitted by Linda Lee Allen. Photographed by Cheryl Cleek.*

The four-crib log barn shown here was built by Robert Harrison Patterson, great-grandfather of the present owner, more than 100 years ago. The Patterson farm is in Perry County's Pineview community. *Submitted by Rex Patterson. Photographed by Wilda Patterson.*

Four cribs in line separated by three runways make for a rare Lincoln County log barn. The 19th-century structure is used for hay storage at present. *Submitted and photographed by Carolyn Hayes Shell.*

Galvanized sheet metal roofing covers John and George Stofel's central aisle log barn, which was built in eastern Maury County above Cedar Creek. *Submitted by Barbara L. Stofel. Photographed by Charles W. Stofel.*

Tobacco dries in the large log double-crib barn situated on the Lora Blevins Farmstead in Scott County near Oneida. The small frame stripping shed attached to one end of the building is a later addition. *Submitted and photographed by Anthony Perkins.*

A drive-in crib barn with a huge hayloft and extensive side shed stands tall in its picturesque setting alongside a country road near Sneedville in Hancock County. *Submitted and photographed by Kathy Robertson.*

A useful side-drive log barn and the land it stands on in Decatur County have been the pride of the Bateman family for many generations. *Submitted and photographed by Ernest D. Bateman.*

At one time, the basic single-pen log corncrib was the most common farm building found in Tennessee. A typical example is located near the Mt. Gilead community in Henderson County. *Submitted and photographed by Ann Reynolds.*

The storage capacity in the loft of this triple drive-in crib has been increased significantly by raising the roof of a historic hewn-log barn located in Knox County. *Submitted and photographed by Roberta M. Felderman.*

When the Davis family bought their Hancock County farm in the 1940s, the old log side-drive crib with an exceptionally large loft already had been there for years. *Submitted and photographed by Jean Davis.*

Initially built as a cantilever barn by carpenter Preston Lloyd in Hamblen County and modified several times since, the yellow poplar and American chestnut log building has seen continuous use since the 1830s. *Submitted and photographed by John Austin.*

Wood Frame Barns

Although log architecture played a dominant role in the built environment of Tennessee during the 19th century, particularly during the pioneer period, the early settlers were quite familiar with other methods of construction. One of these can be described as earthfast, where all of the vertical wooden posts that supported the building were inserted into deep holes dug in the ground. The excavated earth then was shoveled back into the holes and packed tightly around the posts, holding them fast. Modern pole frame structures often employ a similar system.

The wall framing of a building that utilized the most primitive of these earthfast methods, known as puncheon construction, resembled a palisade. Builders dug a trench

Yellow Store Farm in Hawkins County has been in the same family for more than 200 years. Sixth-generation owner Neil Miller continues to use his historic weatherboarded barn today. *Submitted and photographed by Sandi Francis Bolin.*

corresponding to the footprint of the future structure and set a line of vertical posts closely together in it. This row of puncheons supported the framing of the roof.

A more sophisticated earthfast building practice than the puncheon technique was known as post-in-the-ground construction. Carpenters using this procedure set the posts that supported the long walls at regular intervals (usually 10 feet) into well-prepared holes dug about 3 feet deep. The space between two posts supporting a wall is known as a bay. With the posts in place, the structure could be completed using the post-and-beam framing process.

The advantages of earthfast construction were obvious: No foundation meant less building time, more economy, and easy replacement. The main disadvantage, particularly in the South, was that earthfast structures deteriorate quickly in the warm, humid climate. Moisture, rot, and insects take a fearsome toll on wooden buildings in direct contact with the ground. A decomposing structure could cause unforeseen safety and security problems. Nevertheless, over time an uncountable number of earthfast buildings have provided storage and shelter on farms across the state.

Many of the largest and most substantial barns in Tennessee were built using the classic post-and-beam framing process erected on a proper foundation. This ancient method depended on a series of substantial vertical posts that supported a thick horizontal beam. The bottoms of the posts were mortised into a heavy wooden sill borne by either continuous masonry work or a set of rock piers. The tops of the same posts supported the large plate upon which the rafters rested.

In a late 19th-century photograph, several members of the Orr family pose in their Sunday best on their farm near Cleveland in Bradley County. *Submitted by Mellie Brackett. Photographer unknown.*

Dr. William Marion Thompson (second from right, back to camera) practiced veterinary medicine in Crockett County in the early part of the 20th century. The steeply pitched gable-front barn was photographed in 1914. *Submitted by Deborah Heathcott York. Photographer unknown.*

All of the components of the wood frame were hewn, measured carefully, mortise-and-tenoned, and pegged together tightly. Stout 45-degree bracing in both directions kept the structure plumb and true. The resulting rigid framework of interlocked timbers supported a high-quality building that would provide generations of service if protected from hazards. While not common, many of these braced-frame barns still remain on the rural landscape.

In the years following the Civil War, country carpenters started constructing buildings utilizing a new system known as balloon framing. Invented in Chicago in 1832, the technique consisted of fastening milled standard-sized dimension lumber together with machine-cut nails in order to fabricate a lightweight frame that would serve as the structure of the building. The exterior wall and roof surfaces were covered with solid tongue-and-groove sheathing boards that also served as bracing.

The simplicity of constructing the balloon frame made building much easier, quicker, and cheaper. Spindly two-by-fours and two-by-sixes replaced the large timbers of the braced frame; mass-produced nails were substituted for mortises, tenons, and wooden pins; and small groups of farmers, their families, and friends took the places of the itinerant skilled laboring crews. This uncomplicated building system revolutionized the construction of wood frame buildings throughout the United States.

The Anderson family and their descendants have operated Blount County's 500-acre H. F. Anderson Century Farm near Rockford since 1873. The dairy barn was built completely with timber cut off the property. A team of oxen hauled the wood to the site. Note the entry porch on the far left side. *Submitted by Linda Kerr Wells. Photographed by Helen Anderson Kerr.*

Taken in 1917, this picture clearly shows the heavy braced-frame system that supports this massive Blount County dairy barn (also pictured at top right, as it looks today). Master carpenter Bud Chambers and his crew used mortises and tenons to join the timbers together. *Submitted by Linda Kerr Wells. Photograph by H. F. Anderson.*

Jeff Brown's venerable frame barn still stands along Plunkett's Creek in the South Carthage area of Smith County. *Submitted and photographed by Amy D. Waggoner.*

Taken down in 2008, time finally won out over this old barn formerly located on Big Bigby Creek near the Sandy Hook community in Maury County. Note the rock foundation, latticework, and Gothic-influenced openings. *Courtesy of the MTSU Center for Historic Preservation.*

About 150 years ago, William J. Fowler, great-grandfather of William LeNoir, the present owner, began work on his Monroe County stock barn. The louvered blinds add to the character of the well-built post-and-beam structure. The Fowler-LeNoir Farm, established in 1826, is a Tennessee Century Farm. *Submitted and photographed by William C. LeNoir.*

This is a modern photograph of the barn that Louisa Pope and her husband constructed in the late 19th century. Five generations of Pope family women have lived on this Williamson County property over the years. *Submitted by A. B. and Sarah Hicklen. Photographed by Jenn Duncan.*

Louisa Pope and her husband, J. K. Pope, built the farmhouse and chicken coop shown in the background as well as other buildings on their Nolensville farm in Williamson County. *Submitted by A. B. and Sarah Hicklen. Photographer unknown.*

Iron Industry Barns

Middle Tennessee's charcoal iron industry began at Cumberland Furnace in 1796 and spread throughout 15 counties along the Western Highland Rim. In this thinly populated area, the four components necessary for the production of pig iron (ore, limestone for fluxing material, hardwood trees for making charcoal, and water power) could be found in abundance. In the first half of the 19th century, iron smelting took place on vast, self-sufficient plantations, away from population centers.

A furnace in blast required huge amounts of raw materials and fuel, and the people and animals that did the work also had to be sustained. Besides overseeing vast tracts of timber, the ironmasters set aside land for gardens, orchards, fields, and pastures for the support of man and beast. The needs of furnace and farm had to be balanced adroitly in order to provide the necessary workforce for the right task at the proper time. During the first half of the 19th century, the pairing of agriculture and iron-making was a symbiotic arrangement with mutually beneficial effects.

Samuel Stacker and his brother came south from Pennsylvania and began to establish and operate a number of ironworks in Montgomery and Stewart counties. Each of these operations needed a nearby farm to supply the required provisions. Field hands planted crops such as corn, oats, potatoes, clover, and hay and maintained peach and other orchards. Enslaved carpenters constructed all kinds of specialized buildings (corncrib, potato house, ice house, etc.) to process and store the produce.

In addition, Stacker raised chickens, hogs, horses, mules, and cattle that needed proper shelter, so he built individual structures (henhouse, stables, etc.)

Left, farmhands transfer a large stack of hay to the loft in a 1915 photograph. The large post-and-beam barn was built more than 150 years ago by the enslaved workers of Stewart County ironmaster Samuel Stacker. *Submitted by James Brigham. Photographer unknown.*

Above, three generations of the Brigham family pose in front of the historic Stacker barn on Evergreen Farm in 2007. Sheet metal panels replaced the original board roof in 1950. *Submitted by James Brigham. Photographed by Jane Link.*

to protect the animals. In the late 1850s, his slaves erected a 40-by-80-foot hewn post-and-beam barn for use as a cow shed. The ample building provided shelter for the herd at ground level and stored its hay in the loft.

An unusual feature of the barn is the engaged full-length side shed supported by a row of thick wooden columns. A lean-to attached to the main barn along one or more of the sides is commonly encountered on Tennessee farms, but this partially-enclosed one-story addition faintly echoes the posted forebay barns of Stacker's home state. Perhaps it served as a visual reminder of the land he left behind.

Right, Marvin Walker closes the door to the historic log barn located on the old Saline Furnace plantation in Stewart County. The structure was built in 1853 to help maintain the workforce of the ironworks, which manufactured cast-iron sugar kettles for out-of-state customers. *Submitted by Marvin E. Walker. Photographed by Kim Schmidt.*

Stone Barns

The oldest and most durable building material is rock, or stone as it is called when shaped or finished. Stonework represents the most labor-intensive form of construction because each piece must be handled several times before it is finally put in place. The heavy mass of each individual rock aids in resisting movement, and the pleasing natural look of stone allows it to blend with other building materials in an attractive way.

Although the material itself can be found either on or just below the surface of the ground, the gathering, transporting, sorting, and laying combine to make stone the most expensive building method. Stone walls, which must be laid thick enough in order to stand vertically, require an enormous amount of material. Although some kinds of stonework can be dry-stacked (most often fences and fireplaces), the early masons laid the majority of traditional structures in a soft lime or clay-based mortar.

The most common types of rock used for construction purposes in Tennessee are the limestones and the dense sandstones. Both kinds are strong, yet relatively easy to work. The rock can be used either straight from the fields and creek beds or by cutting and roughly shaping them with simple iron tools. The considerable weight and awkward sizes of this basic substance require that those working with it exercise caution at all times.

This limestone apple barn, which also kept onions and potatoes fresh as long as possible, is situated on the Young Acres Century Farm in the Bellwood community of Wilson County. *Submitted by Isabel Young Hall. Photographed by Allen Vance.*

A historic stone dairy barn that stands south of Belleville in Bedford County has been expanded with a wood plank-clad feed room on one side and a concrete block milking room on the other. *Submitted and photographed by Janice McCartney.*

Stone is an uncommon construction material in the agricultural districts of Tennessee. The trees that needed to be cleared in order to raise crops provided plenty of easily-worked timber to fabricate into houses, barns, and outbuildings. Unless the desired rock was close to the building site, a large-scale project would be prohibitively expensive. Despite these obstacles, some farmers still wanted the appearance, security, and permanence that a masonry structure would provide.

Using stone to build stock barns was a wise decision. The solid walls greatly reduced the risk of fire, protected the animals from all kinds of predators, and made it almost impossible for the domesticated beasts to break out and wander off. Although a poor insulator, the tightly-laid stone courses contained and reflected the animals' own heat in the winter months, lessening the effects of severe weather on them.

Farmers who raised and traded horses, mules, and oxen had large sums of money tied up in their herds that often were concentrated in one place. These rural businessmen needed permanent and substantial stables to protect their investment. A stone building met these requirements. Although 19th-century stone barns appeared to be plain, no-nonsense buildings, the choice of masonry identifies them as first-class structures. Although dressed stone often was thought to be reserved for building the main residence of an upper-class planter, other more mundane uses such as sheltering valuable work animals also can be documented.

Wanda Shanks' large stone cattle barn is situated on Big Indian Creek on the family farm in Putnam County. Legend has it that during the Civil War, Confederate cavalrymen used the 1830s structure for sheltering their horses and mules as well as for storing gunpowder. *Submitted and photographed by Wayne Shanks.*

After his distinguished service during World War I, Sergeant Alvin York returned home to the Wolf River Valley near Pall Mall in Fentress County, where he built a handsome wood barn and stone silo. *Submitted and photographed by Tommy O. Jackson.*

Located at Rock Rest Farm along the old Nashville to Louisville Highway in Robertson County, this former relay station for stagecoach horses was placed upon a stone foundation. Note the built-up ramp at the middle of the left side. *Submitted and photographed by Allen Haynes.*

An impressive late 19th-century limestone barn with a monitor roof and ventilators is located on the Wilson Farm near Hendersonville. It is one of the first places in Sumner County where Tennessee Walking Horses were bred. *Submitted and photographed by Bonnie Carter.*

In 1896, the Conger brothers of Beechlawn Farm built a 60-by-100-foot stock barn out of stone quarried from their own land in the Mimosa community of Lincoln County. This Tennessee Century Farm was founded in 1808. *Submitted by Jane Conger. Photographed by Lane Conger.*

Pole barns

Veterans returning home to Tennessee after World War II encountered an agricultural world in transition. While the old ways of working the land, tending the stock, and constructing the farm buildings still remained fresh in the minds of many of the people living in rural areas, it was plain to others that a new day had dawned. On hundreds of farms throughout the state, those who had remained on the home front had realized that when thousands of young men and women departed for the armed forces, much of the work they left behind still had to be done. With little extra help in sight, labor-saving methods for farm tasks proliferated, and new techniques, equipment, and structures emerged to facilitate their spread and use.

The postwar period saw the increasing appearance of large, multi-purpose shelters known as pole barns in every section of the state. Much lower and wider than traditional farm structures, these economical and uncomplicated buildings were designed utilizing standardized measurements (usually in multiples of 4 feet) and manufactured building products such as sheet metal panels, low-pitched wooden or steel trusses, planed precut dimension lumber, and pressure-treated wooden poles.

Inserting the rot-resistant supporting posts deep in the ground and tamping the dirt tightly around them made a standard foundation unnecessary and saved considerable expense. Floors consisted of concrete or packed dirt. Lightweight kiln-dried softwood lumber cut and reduced to uniform specifications made construction easier and quicker than previously had been possible with locally cut and milled hardwoods.

David and Betty Renegar's attractive Bedford County pole barn, right, has the side shielded from the weather left open. The five bays provide for animal shelter, feed storage, and equipment protection. *Submitted and photographed by Holly Gardner.*

Spring Haven Farm in Montgomery County, facing page, is the home of Robert and Dianne Harris' 35-year-old pole barn. They stored hay and tobacco in it for years and still keep horses and equipment inside. *Submitted and photographed by Dianne Harris.*

The large-span roof trusses eliminated the need for partitions, allowing a farmer the freedom to arrange the internal spaces in any manner desired. Long sheets of galvanized steel or aluminum roofing nailed to wooden laths covered the top and sides of the structure. Often translucent panels that matched the pattern of the metal were substituted at regular intervals to provide additional light to the interior.

In addition to a low initial cost, the use of pole barns continued to save owners money on labor by permitting tractors, other vehicles, and machines maximum access to the heavy round bales of hay stored on the ground instead of in lofts. The sizable area under roof also could provide emergency shelter for large numbers of cattle (or other animals) during severe weather. Although possibly lacking in aesthetics to some, the modern pole barn delivers large-scale shelter and storage capacity at the lowest cost.

A pair of open pole barns has the capacity to store large quantities of rolled hay on Windy Hill Farm in the Piney community of Loudon County. *Submitted and photographed by Jaime L. Hull.*

The large traditional wooden barn in the foreground was built around an early 19th-century log corncrib. Next to it is a modern steel-clad pole barn. They are both in use today on the Harris Place Century Farm in Hardeman County, home to the family since 1825. *Submitted and photographed by Darlene Shelton.*

Large rolled hay bales are protected by a well-braced Smith County pole barn. The open arrangement allows the tractor plenty of room to maneuver. *Submitted and photographed by Allen Haynes.*

A 1912 central aisle barn and a more recent open pole barn set in the rural Giles County landscape offer a contrast between yesterday's and today's methods of hay storage.
Submitted and photographed by Jodie Wynne.

University of Tennessee Extension Barn Plans

Established in 1914, the University of Tennessee (UT) Extension is the off-campus educational unit of the university's Institute of Agriculture. This important service ensures that every county in the state has an Extension agent who provides educational programs, up-to-date agricultural information, and household management advice to local farm families. The stated mission of the Extension is "to help people improve their lives through an educational process that uses scientific knowledge to address issues and needs."

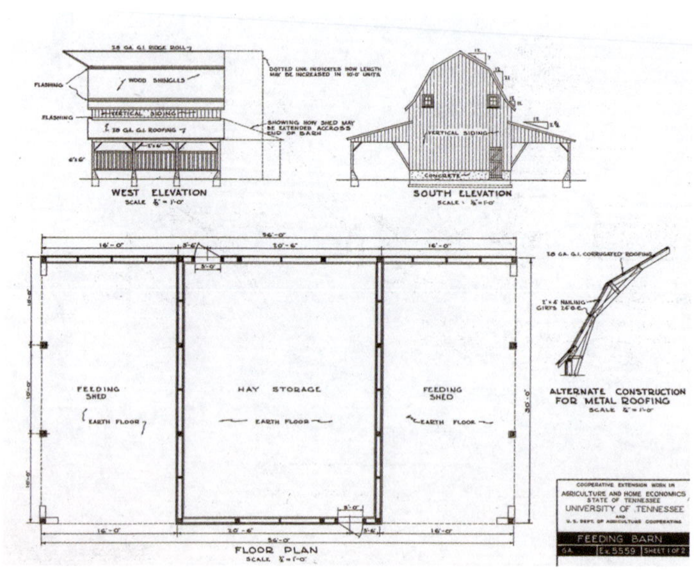

University of Tennessee Extension plan No. 5559 provides complete construction information and specifications for a well-designed feeding barn that remains popular at this time. *Courtesy of the University of Tennessee Extension.*

A revolution in agriculture swept across Tennessee after the end of World War II. The pre-war switch from traditional crops like cotton and tobacco to soybean and livestock production accelerated. In addition, the increased use of tractors and self-propelled farm machinery lessened the need for work animals and expensive extra hands during peak labor periods. Many people from the rural districts headed for urban areas, where new opportunities awaited in factories and offices.

Despite this postwar transformation, farming and living on the land continued to be vital to rural residents who wished to retain the values and other positive elements of their former lifestyle while dealing with the realities of the modern world. To deal with this situation, many people commuted to daily jobs and returned home to their farm-related chores. Although the scope of the overall operations was scaled down in many cases, the necessity and complexity of many agricultural tasks remained.

To assist farmers in their efforts to be more efficient and economical, the University of Tennessee Extension developed, accumulated, and continues to maintain a file of building and equipment plans in cooperation with the United States Department of Agriculture (USDA) and the Cooperative Farm Building Plan Exchange. The Plan Exchange no longer exists, but the blueprints remain in demand. Many of these plans originated at other land grant institutions across the country where they were tested and refined. A list of the designs can be found at the Extension's Web site (see bibliography), and many of these plans can be viewed online and downloaded if desired. The detailed blueprints are provided to Tennesseans at no cost.

The availability of cost-effective, pre-engineered plans has helped both full- and part-time farmers to weigh the available options and make an informed choice before investing time and money in new structures. Each plan takes advantage of standardized lumber sizes and allows for custom additions to suit specific circumstances. The research-based information supplied by the UT Extension has led to widespread knowledge among people in rural areas throughout the state about modern agricultural practices, consumer sciences, and resource development.

Small barns to shelter horses and riding equipment continue to be built in the countryside. Carpenter Teddy Hutchison built a general purpose barn in 1995 for Kevin and Olivia Wann in Stewart County. *Submitted and photographed by Olivia Wann.*

Shelby County is the home of a tidy gambrel-roofed barn with flanking side sheds. This barn was built in 1939 by three brothers: Faires, Harry, and A. K. McCalla. *Submitted by Molly McCalla Hampton. Photographed by Kennon Hampton.*

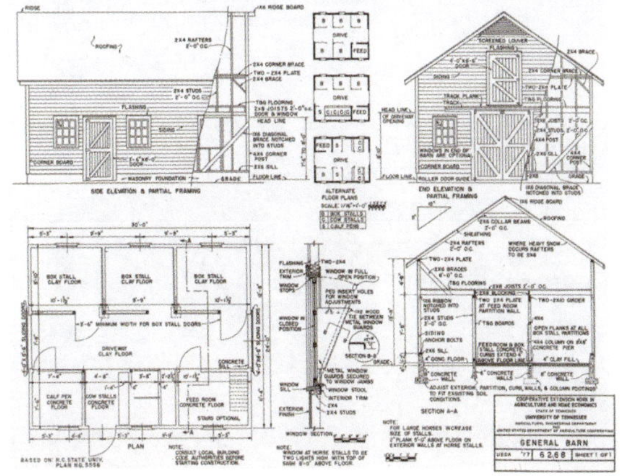

This set of Extension drawings, No. 6268 for a general barn, offers alternate floor plans and other necessary details for a variety of potential uses. *Courtesy of the University of Tennessee Extension.*

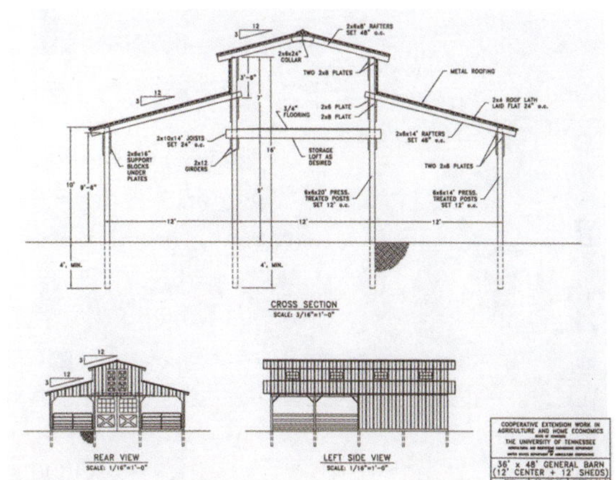

Agricultural blueprint No. T4161, above, illustrates the basic floor plan and specifications necessary to construct a 36-by-48-foot general barn. Note that optional details are provided. *Courtesy of the University of Tennessee Extension.*

Tim and Norma Schvaneveldt's board-and-batten-sided barn, left, seems quite at home in its lovely setting in Woodcock Gulf, Sequatchie County. *Submitted and photographed by Tim Schvaneveldt.*

Ray and Kelly Connor, some relatives, and a family friend teamed up in 2002 to build an efficient wood-frame, sheet metal-clad horse barn in McNairy County. *Submitted and photographed by Kelly Connor.*

Ideal for raising horses, examples of this popular 20th-century farm building type can be found across the state. The barn shown here is located in Anderson County. *Submitted and photographed by Patti Riley.*

Appalachian Barns

After decades of experimentation with compact log or frame structures to shelter foodstuffs, animals, and livestock feed, Tennessee farmers have erected them in an almost bewildering number of forms. Among them is a type called a front-drive crib, consisting of a single corncrib with an extended gable roof held up in the front by posts and accessible from an open aisle running from side to side, perpendicular to the ridge. This efficient arrangement allowed the driver of a wagon to pull into the sheltered area and unload out of the weather.

This vintage photograph depicts an Appalachian barn built by John Sherfey in 1916 on Knob Creek in Washington County. It has a screen wall on the front side of the cross aisle. The building is still in use on Pioneer Homestead, the Holleys' Century Farm. *Submitted by Margaret Sherfey Holley. Photographed by Lelia Sherfey.*

Farmers who liked this handy set-up, yet needed more space, placed another unit next to the existing one and then a common roof over both, thus creating a farm building referred to by some as an Appalachian barn. This barn type has two main varieties. The first consists of two or three cribs side-by-side with a cross aisle in front that services each crib. A thin screen wall on the other side of the aisle serves as the front of the barn, although it may be pierced by some windows and doors or even left open to allow light into the interior. An opening for hay appears at the top of the front gable wall.

The second variation of the barn tends to be more substantial than the basic configuration described above. In this version, the interior crib area is divided in half by a secondary aisle that runs parallel to the ridge and forms a T-shape with the primary aisle that runs across the front. Each of the two main cribs can be divided into stalls that open to the secondary aisle. Small windows on opposite sides of the barn provide light and cross-ventilation.

A narrow corncrib, accessed from the primary aisle, often takes the place of the front screen wall described above. In other examples, the corncrib was added on the outside of the screen wall under its own shed roof. A protective hay hood usually graces the apex of one of the gable ends.

The Appalachian barn is a distinctive farm building that flourishes mainly in the eastern counties of Tennessee, but it can be found in scattered places almost anywhere in the central or southern Appalachian Mountains and as far north as Ohio and Indiana. While somewhat uncommon, it reflects a certain notion about the proper arrangement of interior barn space held by a segment of the farm population.

Ruth Gray Seals poses near Richie Gray's farm buildings in this 1947 photograph taken in Claiborne County. Just behind her stands a typical front-drive crib framed out of sawmill lumber and covered with wooden shingles. The small building can be viewed as the earliest form of the Appalachian barn. *Submitted by Mata Banks. Photographer unknown.*

A large Appalachian barn located near Johnson City in Washington County features a central ventilator, pent roof protecting the entrance, and a small shed addition. *Submitted and photographed by Carolyn Whittenburg.*

Photographed from this angle, it is easy to identify the aisle running from side-to-side at the front of this Blount County barn. Note the roof of the main barn does not extend over the corncrib. *Submitted and photographed by Darrell Watkins.*

Family members proudly stand next to the barn that was built by Mr. and Mrs. Whitelaw Dent Johnson in 1924. The crew repaired the foundation, added new wooden siding, and replaced a deteriorated shed during the hot summer of 2007 in Johnson County. *Submitted by Mr. and Mrs. C. F. Lovell. Photographed by Lois Lovell.*

The lovely setting highlights a more substantial variety of an Appalachian barn featuring a narrow corncrib located at the front and a hay hood above. This well-known structure is situated in Cades Cove in Blount County. *Submitted and photographed by Joe DeLay.*

Bank, Raised, and Foundation Barns

Much of Tennessee's rural landscape that lies outside of the Gulf Coastal Plain can be described as a land of hills and hollows. Farmers intuitively sought the most level terrain for raising crops and often sited their buildings on sloping ground. In order to cope with these circumstances, barn builders sometimes partially excavated the lower story and dug the structure right into the slope, producing a bank (or side-hill) barn. Settlers moving from Pennsylvania brought the idea of the bank barn with them as they traveled southward.

Either the side or the gable end of a bank barn could be built into the hill, which would provide a frost-free manure storage area. The lower story also housed the animals and was accessible only from the downhill side. The walls of this bottom level usually were composed of stone, although many barns were made completely out of wood. In the 20th century, the use of cinder blocks or poured concrete for the foundation walls of these bank barns became increasingly more common. It was important to slope the ground away from the building so that proper drainage ensued.

Wagons entered the main floor directly from the upper level, where they could unload either there or into the loft. In cases where there was a space between the bank and the barn, the carpenters built a bridge to connect the two. The use of a bridge instead of an earthen ramp on the uphill side allowed the builder to place windows in the exposed wall in order to ventilate the ground floor. Among the advantages gained by the use of the bank barn were a warm stable area in winter that was cool in summer, storage for root crops, and additional room to work under the same roof.

A raised barn shares many similarities with a bank barn. The main difference is that a raised barn is built on relatively level ground, and a ramp of earth or wood is constructed from the ground to the main floor to provide access. A special roof structure called an entry porch may be built above the wagon doors to protect the entrance. Of course, the bottom floor can be easily entered from any side desired.

The foundation barn is built on level ground and has no direct exterior wagon entrance to the upper story. A large hay hood often is located at one gable end. Generally, the builders constructed the first-floor walls out of masonry and the second story from wood. Hinged or sliding doors can be found at the center of each end of the lower story. Windows were uncommon on the older buildings, but they tend to appear on more recent barns. Large-scale foundation barns with accompanying silos continue to be built in the Amish and Mennonite communities in some of the more rural areas of the state.

Marion and Bertha Gonce stand in front of their bank barn in this photograph taken in 2004. The couple lived on the Marion County farm since purchasing the property in 1934. *Submitted by Bertha Gonce. Photographed by Leonard Wynne.*

A large foundation barn and silo dominate the landscape of this Amish farmstead in northern Lawrence County. Note the concrete block first story and the lack of wagon access to the second story from the outside. *Submitted and photographed by Diane Tatum.*

After the Turner family's bank barn was framed in by carpenter Fate Allen in 1902, neighbors joined in to help finish it within the week. The building is located on the Steele Century Farm in Dickson County. *Submitted and photographed by Janelle Steele Turner.*

James Gray's gambrel-roofed bank barn in McNairy County provides direct access to one end of the upper floor from the road in front. *Submitted and photographed by Jennifer White.*

The walls of the ground floor of this Humphreys County bank barn are composed of concrete blocks. An earthen ramp was placed at the far gable end for wagon access to the upper level. Henry Parker built the structure in the mid-1960s, and it is now owned by the Hooper family. *Submitted and photographed by Becky Matlock.*

Standing in the middle of a Lawrence County field, a raised wooden barn has the ramp that leads to the main floor in the middle of its long side. Amish carpenters built the structure in 1951. *Courtesy of the MTSU Center for Historic Preservation.*

The Bivens family's 19th-century barn that originally occupied the site in Lincoln County burned and was rebuilt by master carpenter Roscoe Epps in 1936. The entry porch protects the bridge that allows wagons to enter the barn at the second-story level. *Submitted by Pat Bivens. Photographed by Skip Bivens.*

Located on the Miller Century Farm in Dickson County, the bank barn shown here accommodates hogs on the lower level, allows for equipment maintenance and other storage on the second, and holds hay in the loft. It was built by founder Augustus Lincoln Miller, who emigrated from Gotha, Saxony, in the beginning of the 20th century. *Submitted by Debby Stewart. Photographed by Skip Delius.*

Lewis County is home to a number of large gable-entrance bank barns. Structures like this can be traced back to the German-speaking settlers who came to Middle Tennessee during the late 19th century. *Courtesy of the MTSU Center for Historic Preservation.*

Cantilever Barns

Possibly the most unique traditional farm building found in Tennessee is the log crib barn with a cantilevered second story overhead. The large projecting wooden beams that hold up the loft are supported only on one end and bear the load on the other. The basic principle of the cantilever has been known to builders for centuries and can be traced at least as far back as Roman times.

European prototypes influenced American barns, and a considerable number of barn varieties originated in the German-speaking countries. Immigrants who came to the New World modified these older structures to suit the particular conditions that they found when they arrived. It seems likely that the inspiration for cantilever barns came to Tennessee with travelers from the Germanic settlements of Pennsylvania via the Great Valley of Virginia.

The most common form of this unique barn type features two log cribs at ground level and an extended loft overhanging on all four sides. The lower level served primarily as stables for horses and mules as well as for sheltered work areas, while the upper floor stored large quantities of hay. The double-crib is by far the most common variation of cantilevered barn, but examples of single-crib, four-crib, and five-crib barns can be found scattered throughout the Great Smoky Mountains area.

These structures are found within a fairly limited range, concentrated primarily in Blount and Sevier counties, where 90 percent of the known cantilevered barns in the United States exist. These buildings share certain features besides the log crib first story and the cantilevered frame structure above. All rest on rock pier foundations. The logs usually are hewn and notched at the corners, but some rural carpenters did use round logs at times. Several different notches (half-dovetail, V, square) have been employed on these barns, but the half-dovetail type appears most often.

An uncommon single-crib cantilever barn with two open side sheds has been preserved at the Cable farmstead in Cades Cove by the National Park Service. It was built to be bear-proof. The posts are not original. *Submitted and photographed by Tommy O. Jackson.*

Built by George O. Stepp in 1875 in the Range community of Carter County, the Hughes family's late 19th-century barn was utilized for drying tobacco and other agricultural functions. The two rows of posts disguise the signature cantilevers. *Submitted by Ted Hughes. Photographed by Walter G. Martin Jr.*

The spaces between the logs are not chinked, and the cribs have only dirt floors. Many of the complex cantilevered barns originally had a threshing floor in the runway, but 20th-century owners often removed it to allow access for wheeled agricultural equipment. Hard-working craftsmen hewed the massive timbers of the second story and put them together with mortises and tenons tightly fastened with wood pegs. Lofts were sided with either vertical planks or horizontal weatherboards. The once ubiquitous board roofs mostly have been replaced with galvanized sheet metal panels.

It appears that East Tennessee farmers began to build these cantilevered barns in the period prior to the Civil War and continued until just before World War II began. Interestingly, more than half of the surviving barns date their construction to the last 20 years of the 19th century. Changing economic conditions in the postbellum period may have played a role; skilled builders coming south could have been responsible; a rise in the number of local sawmills might have been a factor — any number of possible reasons could account for this short-lived rural building boom.

The cantilever barn is a rare and somewhat unfamiliar building to many people. Its limited range and relatively small numbers have contributed to this lack of knowledge. While individual structures have been identified in several counties scattered across the state, others remain unrecognized. Hopefully, many of those forgotten structures will be discovered and cared for properly so that future generations will have examples of this distinctive folk building to study and appreciate.

A classic Blount County double-crib cantilever barn, located in Cades Cove, is one of many historic buildings maintained by the National Park Service in the Great Smoky Mountains. It is reputed to be the most photographed barn in the United States. *Submitted and photographed by Deb Dadisman.*

The barn pictured here had stood in two different states and three different counties but never had been physically moved (state and county boundary lines were redrawn over the years). It recently was disassembled in Loudon County, transported, and rebuilt at the Great Smoky Mountains Heritage Center in Townsend. *Submitted and photographed by Hank Caruso.*

Five generations of Johnson County's Wills family have used the cantilever barn on the Roan Valley Dairy Century Farm. The chestnut log cribs have housed work horses, milk cows, and sheep, while the loft above has stored hay, wheat, and tobacco. *Submitted by John V. "Slap" Wells. Photographed by Margaret Wills.*

Although only two or three years old in 2007, a charming Blount County structure plainly shows the influence of the traditional cantilever barn on its construction. Owner Ron Campbell built the little barn for storage. *Submitted and photographed by Linda Arwood.*

A good example of a cantilever barn can be found at the Museum of Appalachia in Anderson County. The historic structure is part of an outstanding collection of folk buildings moved to the site and carefully restored by founder John Rice Irwin. *Submitted and photographed by Randy Ball.*

Central Aisle Barns

Perhaps Tennessee's greatest contribution to the built agricultural environment of the South has been the development of a distinctive farm structure called either a central aisle barn, a transverse-crib barn, or, more familiarly, a general livestock barn. The geographical area where the majority of these barns can be found is roughly bordered by the Atlantic Coastal Plain on the east and the Ozark Mountains to the west, and the Gulf Coastal Plain to the south and the Ohio River on the north.

The main characteristics of the multipurpose central aisle barn include a wide runway in the middle of the building that is parallel to the ridgeline of the roof, front and rear-facing gables, a line of separate cribs on each side of the open aisle, and a loft above. Although farmers used the building in many ways over the years, feed storage and animal shelter constituted the primary functions of this barn. It was ideally suited for the state's self-sufficient inhabitants who practiced small-scale general agriculture built around a mixture of livestock raising and grain cultivation.

The origin of the central aisle barn can be traced back to the Watauga district of East Tennessee, late in the 18th century. The first generation's agricultural efforts had been so successful that many post-pioneer farmers turned their thoughts to a more intensive land-use strategy, one that would result in greater production and, consequently, would require bigger buildings.

As described earlier in a previous chapter, farmers created a new, yet familiar, structure with more flexibility and capacity than the kinds they had been using by building a double-crib open runway barn facing a similar one and then placing them both under the same roof. This resulted in a rectangular barn with a crib in each corner and a cross-shaped open aisle separating the four cribs.

It didn't take long for some farmers to realize that only one runway was actually needed, so after blocking off the cross aisle at both ends and reapportioning the newly freed-up space for additional cribs, a barn was born. The gable-to-gable runway proved to be the better choice than the earlier configuration because it resulted in much more headroom in the loft, and the structure itself could be easily expanded lengthwise as much as necessary. On the average, central aisle barns have from two to six individual cribs on each side of the runway.

Although the early central aisle barns were made of notched logs, the vast majority of the buildings that exist today are balloon-framed with standard-sized dimension lumber and clad with either horizontal or vertical wood planks. Frame construction is considerably more versatile than log, brick, or

The heart of Shelby County's last dairy farm still beats faintly in this recent picture. From 80 to 100 registered Holsteins once called the barn home. Three generations of the Moffatt family continue to live on the land. *Submitted and photographed by Sandra Hill Jolly.*

An old photograph depicts a well-designed central aisle barn situated on the Lewis farm in Sumner County. Built in 1937, the structure is still in use and in excellent condition. *Submitted by Mary Ellen Lewis Glover. Photographed by Mrs. Reno Lewis.*

stone, and this flexibility allowed for increased complexity in the arrangement of interior space. In addition, either open or closed sheds could be added to the sides of the structure, which would significantly increase its size. The sheds themselves could be used for additional runways, with the stalls off to one side or the other.

Log, stone, and brick barns rarely were painted, but wood-sided buildings were different. Besides improving the appearance of a structure, painting protected the wood and greatly extended its life. Some farmers made their own paint from a milk base tinted with naturally-occurring compounds such as iron oxide, but the growth of industrialization led to widespread availability of cheap manufactured materials.

Painted Tennessee farm buildings are usually colored red or white. These coatings consisted of either red or white lead pigments mixed with a linseed oil binder. Red lead was by far the cheaper ingredient and lasted much longer, which is probably why there are so many red barns on the landscape today. Some thrifty farmers pocketed extra cash and a free paint job for at least one side of their barns in exchange for the use of the building as a billboard. (These advertising barns are discussed more fully in Part III.)

The basics of the central aisle barn have been described here briefly, but we've by no means fully explained the possible variations in size, details, or appearance. Different materials, roof types, additions, door and window placements, ventilators, stanchions, and other characteristics have been blended in myriad combinations that continue to attract our interest today. The almost limitless arrangements available to farmers demonstrate the building's adaptability and continuing popularity throughout the state. Without question, the frame central aisle barn is the most common type found in Tennessee, and readers will find a number of images of it throughout this book.

Plenty of hay can be stored in the loft of this early Bedford County central aisle barn. Constructed with logs in the early part of the 19th century, the impressive gambrel roof is a later improvement. A local blacksmith made shoes for the horses of both armies here during the Civil War. *Submitted and photographed by Colleen Adcock.*

In 1917, skilled German immigrants built a massive brick dairy barn on Grandview Farm in the Mimosa community of Lincoln County. In addition to cattle, the structure has housed mules, sheep, goats, and Tennessee Walking Horses. *Submitted and photographed by Melanie Goodin.*

In its heyday, this barn was at the center of activity on the 700-acre farm of Joe Sutherland near Mountain City. An original section of the barn is more than 100 years old. *Submitted by Hill Ward. Photographed by Robin Conover.*

The Hunter brothers' cattle farm on Anes Station Road in Marshall County utilizes this large central aisle barn for many farm functions. It was constructed early in the 20th century. *Submitted and photographed by Sherry Buchheit-Hunter.*

Albert Lee Webb's multipurpose farm building is located near Niota in McMinn County. Erected by Gus Roberts, it is one of the older barns known to exist in the community. *Submitted and photographed by Sharon Webb.*

Originally built for the McWilliams family about 1920, this large barn is constructed entirely from heart pine. It is located just outside of Dunlap in Sequatchie County. *Submitted by Jack H. Johnson. Photographed by Norma Johnson.*

J. D. Beene's big red hay barn stands near Graysville in Bledsoe County. Carpenter Tom Fuller built the structure in 1940. *Submitted by J. D. Beene. Photographed by Betty Beene.*

Located on the Price Brothers Farm near Lynchburg in Moore County, a gambrel-roofed central aisle barn with a shed on each side provides plenty of storage space. *Submitted by Joel and Sarah Sanders. Photographed by Scott Newman.*

One of the better-known Middle Tennessee barns stands at the entrance to Rutledge Falls near Tullahoma in Coffee County. Built during the 1920s, the attractive structure is a favorite with photographers year-round. *Submitted by Susan Jennings Owens. Photographed by S. Dean Northcutt.*

English Barns

The settlers who came from the British Isles and other places in Europe brought the folk ways of their native lands with them to the New World. Part of this cultural memory included mental templates of useful structures including barns and outbuildings. These new American farmers modified their time-honored designs in response to the unfamiliar conditions that they found in North America.

Although the original purpose of Buford and Jenny Hollis' Montgomery County barn, above, was to store hay, they have used the structure more recently to dry burley tobacco. *Submitted and photographed by Linda Davis.*

The traditional English barn was a simple one-story, side-gabled building composed of three parts: an open central threshing floor flanked by two equally-sized storage areas. Because of the milder climate in the United Kingdom, this one-story structure with a loft was used only for processing and warehousing grain, not housing livestock. In America, with time and labor short, creative agriculturists saw the advantages of combining different farm functions under the same roof and began to put animals, feed, tools, and equipment into different sections of the same building.

Although the building was similar in outward appearance to the Old World model, the interior of the American version was partitioned so that it could house a number of separate farm activities. When threshing machines became common by the 1850s, the outdated wooden floors were pulled up and the space used as a runway. By redefining familiar spaces within an old form, adaptive farmers created a true multipurpose agricultural building that fulfilled their current needs in an economical way. This tendency to adapt traditional methods to solve fresh problems became part of the Americanization process.

Barns built of stone were the norm in England, but on this side of the Atlantic Ocean, timber ruled. Carpenters often used the post-and-beam technique of construction on a continuous stone foundation to erect these barns, but as time went on, separate rock piers would suffice to support the building. Heavy vertical planks nailed to the framing and long board shingles covering the roof completed the structure. Windows were rarely

Right, the first owner of this Lawrence County barn, Marvin Phillips, never drove an automobile and instead used a wagon and team to haul farm products. His brother, Wendell, ran a blacksmith shop in the lean-to addition on the downhill side. *Submitted by Kathryn Roberson. Photograph courtesy of Elaine Pennington, photographer unknown.*

Behind the split-rail fence is a wood-sided English barn with a shed addition on one side. It is located on Campbell Road in Johnson County. *Submitted and photographed by Annlynn Ward.*

needed on these barns; light and air would filter through the cracks between the siding boards.

The entry to an English barn was through the doors centered on each of the long walls. Originally, these doors were placed directly opposite of each other so that the prevailing winds could flow through the building and carry off the chaff of the threshed grain. After threshing floors became obsolete, this space became a runway that allowed a team pulling a wagon to enter and leave the barn without turning around.

English barns are rare in Tennessee today. Much more common are the log crib or frame barns with their main entrances on the gable walls rather than in the middle of one of the long sides. Nevertheless, this venerable farm structure has a long and useful history as part of the Volunteer State's agricultural past.

The hand-hewn beams of Charles and Marie Keys' barn were put together using wooden pegs. Built just after the Civil War, the Washington County structure has long been called "The Old Barn." It is part of the Brookside Century Farm, established in 1793, one of several certified family farms founded before Tennessee's statehood. *Submitted and photographed by Marie E. Keys.*

Vernon Shell built a gambrel-roofed English barn on Morgan's Branch Road in Carter County during the 1950s. He installed the sheathing diagonally for bracing purposes. *Submitted by Ruby Shell. Photographed by Charles G. Woods.*

Four generations of Lauderdale County's Cates family have used their English barn to house animals and store equipment. Built in the 1920s, it remains in use. *Submitted and photographed by Amy J. Eads.*

Gambrel-Roof Barns

The brick building in the foreground once housed a creamery operated by college students attending classes in Harrogate to defray the costs of their tuition. Built in 1915, it is now the home of the School of Veterinary Technology at Claiborne County's well-known Lincoln Memorial University. *Submitted and photographed by William H. P. Nevils.*

The gambrel roof originated in Europe and was brought to America by several different ethnic groups as part of their heritage. Although built throughout the Northeastern United States since Colonial times, it did not find widespread acceptance in the South and Midwest until after the Civil War. During the 20th century, the use of this roofing system on farm buildings of all kinds rapidly increased, and before long the gambrel roof became the method of choice for covering central aisle stock barns.

The development of the gambrel (or curb) roof was a result of the constant search for more efficiency. The word "gambrel" is derived from the hock (the bent part) of a horse's leg, which is also known as a gambrel, and soon came to describe the bent rafter line of a building's roof, compared to the straight line of the familiar gable. Of course, these long rafters were not really bent; rather, a full-length timber was replaced by two shorter rafters, each one cut to a different pitch and then joined together, forming an angle.

The longer rafters with the steeper pitch usually appeared at the lower side of the roof, resulting in a pleasing bell-like shape. The use of the gambrel roof not only allowed the use of shorter (thus stronger) timbers, it also increased the volume of space in the loft considerably without raising the height of the side walls. When the paired rafters were fastened to their respective top plates and joined at the ridge, the result was a self-supporting roof with no internal obstructions to interfere with the movement of hay.

Innovative barn-builder Edwin Roscoe Epps Sr. pioneered the use of self-supporting roofs in southern Middle Tennessee beginning in the mid-1930s. The owner of a sawmill located in Epps Hollow near Belleville in Lincoln County, he supplied the lumber and crew for constructing barns and houses both locally and in the surrounding counties for nearly 25 years. A good number of these well-built structures remain in use on the landscape today.

Edwin Roscoe "Pa" Epps (1890-1964), Lincoln County's master barn builder. *Submitted by Edwin R. Epps Jr. Photographer unknown.*

The economy of the gambrel roof was hard to beat. Its efficient design produced more loft capacity with less framing and sheathing needed to do the job than an ordinary gable roof required. With all of the livestock at ground level, hay could be dropped directly into their mangers, saving valuable time. For many reasons, the popularity of the gambrel-roof barn remains high to this day throughout Tennessee.

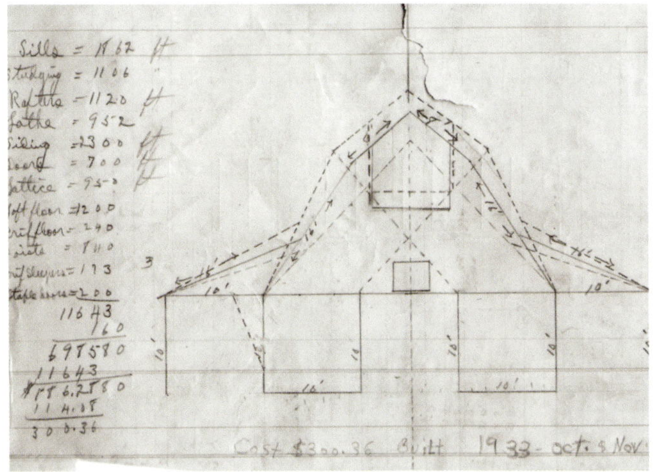

Reproduced here is the original plan used to construct Bodie Jonakin's gambrel-roof barn in Obion County in the autumn of 1933 at a total cost of just over $300. It took almost 12,000 board feet of lumber to do the job. *Submitted by Christine Laird. Plan drawn by Bodie Jonakin.*

Master builder Roscoe Epps constructed the symmetrical gambrel-roof barn near Fayetteville in Lincoln County in 1956. Owner Don Wilkes is a retired Williams Lumber Company carpenter. *Submitted by Don and Nell Wilkes. Photographed by Barbara Ann Wilkes Groce.*

Located on White Ridge Road in Grundy County, this gambrel-roof barn was built by the Charlie White family in 1919. Charlie's grandson, Jack, now owns the place, which is certified as a Century Farm. *Submitted by Jack White. Photographed by Janice White.*

In this vintage photograph, Lucille Parks Bones and her son, Horace Parks, are shown in front of the multipurpose barn on the family homestead. Her parents, Mr. and Mrs. Tom Parks, moved onto their Lincoln County farm in 1940. *Submitted by Odey C. Parks. Photographed by Barbara Moore.*

The main part of an endangered Lake County gambrel-roof barn was constructed some sixty years ago in the Madie community. A street in a neighboring subdivision can be viewed through the large opening. *Submitted and photographed by John E. Dial.*

In 1957, Dave Hestand built a gambrel-roof barn in the Moss community of Clay County and hosted rodeos, bull-riding contests, and even the local fair. Current owner Jerry McLerran uses the building to shelter his cattle and dry tobacco. *Submitted and photographed by Kristi Shirah.*

Everette Holland built a small dairy barn on his land near Sparta in White County in 1959. Although the farm has been sold to new owners, granddaughter Abby Hyde posed happily under the family surname painted at the top of the gable. *Submitted and photographed by Elaine Holland Feltner.*

The builder of a neat Cocke County barn placed slats just below the eaves and overhangs for ventilation. The well-maintained building is located in the Point Pleasant community. *Submitted and photographed by Stephanie Holbert.*

Carroll County's William Bodkin Ford built his dairy barn in 1948. The original structure measured 50 by 60 feet and stored corn and hay besides sheltering the herd. *Submitted and photographed by Olivia Blazer Bowen.*

Gothic-Roof Barns

Presenting a pleasing silhouette against the distant sky, the Gothic-roof barn represents one of the last of the handcrafted building types. The graceful curves of the paired rafters that form the distinctive arched roof mask the great strength necessary to resist the stresses placed upon the structure while simultaneously inspiring onlookers with their sublime beauty. This aesthetic appeal has blended with the necessity and practicality of maximizing the available space in the loft to produce an attractive farm building that met the requirements of late 19th-century scientific and large-scale commercial agriculture.

The origin of Gothic (also known as round, arched, or rainbow) barns seems to be in Michigan, where the first known prototypes appeared about 1885. Because each side of a Gothic roof is part of a perfect circle, the earliest rafters used were fashioned from short pieces of 10-inch-wide boards sawed to the proper radius and then nailed together (with the joints staggered) to the specified thickness. In order to prepare these rafters easily, access to a power saw was essential.

Within a decade, carpenters had developed a simpler way to fabricate Gothic rafters. This method entailed laying out the rafter pattern on the loft floor of the unfinished barn, nailing wood blocks on each side of the pattern to form a jig, bending several thicknesses of three- or four-inch-wide planks between the blocks, and fastening them securely to each other to form a multi-layered laminated rafter. The completed piece (known as a bent) was removed from the jig, set aside carefully, and the process repeated until all of them were finished. Then the completed bents were erected in sequence.

Despite the simplicity and usefulness of these laminated bents, widespread utilization of this technology spread slowly. However, continuous experimentation at land-grant universities and other agricultural research facilities resulted in the publication of nationally available plans that helped to popularize the advantages of the Gothic barn. Large manufacturing concerns such as Sears, Roebuck, and Company produced complete building packages that provided precut materials and full instructions for siting and constructing these barns. By the 1940s, theory and experience had combined to establish the Gothic-roof barn as the centerpiece of many Tennessee farms.

Progress almost claimed the 1950 McDuffee dairy barn in Sumner County until developer Teddy Adams decided to preserve the building as a piece of local history. Craftsmen Will Cantrell and John Elizer restored the structure, which marks the entrance of the Somerset Downs subdivision. *Submitted by Thomas Newton. Photographed by Joel Thomas.*

The snow-capped rolls of hay seem to be somewhat forlorn as they rest in front of the empty loft of Will Sewell's Putnam County Gothic-roof barn. The change from rectangular bales to round ones spelled the end for many traditional farm buildings. *Submitted and photographed by Carolyn Whittenburg.*

When John E. White's family constructed their barn near Rogersville in Hawkins County during the summer of 1955, even the grandchildren pitched in to help. *Submitted and photographed by Abby Gail White.*

Although at a distance a large Washington County farm building appears to be made of stone, in reality it is composed of three-inch-high concrete blocks. Constructed in 1929, the 30-stall dairy barn is on the campus of historic Washington College Academy in Limestone. *Submitted and photographed by Nancy Adgent.*

Owner John M. Webb had to overcome his Chester County neighbors' skepticism when he built a Gothic-roof structure of his own design in the early 1950s. The sheet metal roofing panels fit together fine, however, and the building stood until a windstorm blew it down a few years ago. *Submitted by James A. Webb. Photographed by Janeane Moore.*

These two Marshall County farm buildings — the round-roof wooden barn on the left and the gable-roof metal pole barn to the right — offer a stark contrast between the old and the new. *Submitted and photographed by Margie Baker.*

The timber for this 60-year-old barn was cut right off the farm in Henderson County. The owners, Ernest and Rosa Mae Scott, hauled the logs to the sawmill with a wagon and team of mules. The roof attracted lots of attention when it was constructed. Today the landmark "Round Top Barn" is home to goats, sheep, donkeys, and a llama. *Submitted and photographed by Marie Scott Wharton.*

Several generations of the Qualls family pose proudly in front of a Sumner County Gothic-roof barn. The large barn, built by their relative Tom Stone in the early 1940s, remains in the Stone family today. Note the large hay hood at the gable and silo to the rear. *Submitted by Dan J. Qualls. Photographed by Benita Qualls.*

Although not technically a gothic-roof barn, an imposing DeKalb County farm building strongly resembles one. The builder utilized three different roof pitches on each side to approximate the shape of curved bents instead of cutting the rafters to the desired circumference. *Submitted and photographed by Diane Tatum.*

A large multipurpose barn near Johnson City in Washington County.
Photographed by Robin Conover.

Part II

Barns As Shelters and Storehouses

Introduction

As discussed and illustrated in the preceding chapters, even before Tennessee became the 16th state, barns of logs, timber, and stone were built to shelter livestock and store corn, grain, and hay. Later barns were constructed to specific and more sophisticated designs for particular needs — dairy, tobacco, sheep, or goats, for example. Because barns are the focal point of the farming operation, most are used for a variety of purposes. As farmers diversified for economic or practical reasons and as new technologies, methods, and equipment changed the agricultural composition of farms, barns were adapted and altered to suit each generation's needs. Over time, sheds were added for tractors, machinery, and tools; sections were reworked for different animals and products; and provisions were made for the workspace needed for a variety of tasks. Barns have always been storage facilities for all sorts of things needed to operate a farm and for those items that might possibly be put to good use again one day. Part II focuses on barns as accommodation for animals and crops.

"Perhaps another reason barns are respected and revered in our memory and our art and photographs is that they represent work done and the harvest in it. The stored fodder and corralled livestock give the farmer a sense of accomplishment and security. At the end of the day, there is something tangible to see. I think they're picturesque because they have simple lines that seem to blend with the landscape. The rural background often is a masterpiece of beauty itself. However, the barn is still the focal point, the center of interest in most scenes. And both new barns and old barns draw the photographer's eye."

From *Why Barns?* Submitted and written by Anne C. Sowell of Hendersonville. (Sumner County)

Right, John Kemmer's barn in Grassy Cove near Spring City in Cumberland County. *Photographed by Robin Conover.*

Horses

Whether bred for the saddle, harness, racing, jumping, pleasure riding, show, or work, horses have long occupied the premier place on the farms of Tennessee. By 1805, Andrew Jackson, the seventh president, was part owner of a racetrack at Cloverbottom (Davidson County) and emerged as the leading horse breeder and racer in the early years of statehood. Tennessee, along with Kentucky, was considered the center for horse breeding and racing in the United States for most of the 19th century. Sumner County supplied many of the South's finest race horses, and the Belle Meade stables were known nationally and internationally for their thoroughbred stock. Race tracks and jockey clubs were active in many counties by the 1840s. Horses worked side by side with people both on the farm and in towns in many capacities, provided transportation in a variety of drawn vehicles or on their backs, and made life easier and richer in many ways.

Russell Duvall, center, has lived most of his life on the family farm in Lauderdale County. Built in the late 1800s by Buck Griffin, the barn has been home to cows, pigs, goats, dogs, cats, and mules as well as horses. With Russell is his son, Roger, grandson, Tyler, and Princess. *Submitted and photographed by Connie Duvall.*

In the first decades of the 20th century, horses continued to provide transportation and power for clearing and tilling the soil as well as harvesting crops. As automobiles, mechanized machinery, and electricity became available and affordable, horse power ceased to be a necessity. Tennesseans, however, continued to ride, raise, show, and work horses of varying breeds. The Tennessee Walking Horse, originating from the bloodlines of many breeds including the Naragansett Pacer, the Morgan, and the Standardbred, evolved in the first half of the 20th century into one of the most beautiful, exciting, and recognizable equine performers in the world. In 2004, the U.S. Department of Agriculture reported that the state had 61,000 Tennessee Walking Horses, 47,000 Quarter Horses, 9,100 Spotted Saddle Horses, 6,100 American Standards, and 65,000 other breeds. Currently, Tennessee's equine population is the second largest in the United States, and the value of the state's horses is estimated at more than $565 million.

Built from logs, milled lumber, brick, and stone, horse stables are found in general livestock barns of many sizes and types and in impressive facilities constructed especially for their breeding, training, and comfort as well as trading and sales.

Horses, whatever their breed and whatever their purpose, have always been given a special place by Tennesseans in their history, on their farms, and in their barns.

When Philip Jones received this photograph from Winnell Pittenger, his 90-year-old aunt, she had identified every person at this Hillsboro farm in Coffee County. The picture was made to celebrate the completion of the horse barn and documents not only the style and craftsmanship of the barn but the family's saddle horses and horse-drawn carriage as well as a relatively new form of transportation — the bicycle. *Submitted by Philip Jones. Photographer unknown.*

Left, Joe Haley Young was showing a young gaited horse to potential buyers in 1921 when this picture was taken. On Young Acres, now a Century Farm in Wilson County, Joe and his father, Calhoun Young, who built this barn in 1903, worked together training and showing American Saddle Bred horses. They pioneered the cutting of the tails of gaited horses to give them the distinctive look they display when trotting. Isabel, daughter of Joe Haley Young; her husband, Thomas Hall; and their two children are well-known riders and breeders of American Quarter Horses. *Submitted by Isabel Young Hall. Photographer unknown.*

A yearling filly, left, looks on as horses are being shod inside the Marshall County barn owned by Joel and Patti Musgrave. *Submitted and photographed by Jann Kennedy Beasley.*

"Sonny" waits to be fed in front of the 1890s barn owned by Mrs. Baker in Dyer County. *Submitted by Mrs. Guliford Baker. Photographed by Lee Barner.*

The Milky Way Farm, right, was developed in the early 1930s by Franklin Mars, founder of Mars Candies, Inc., and his wife, Ethel. During the Great Depression, the Giles County farm was the largest employer in the county. Barns for horses, cattle, and sheep were part of the 2,800-acre farm. *Submitted and photographed by Lisa Tidwell.*

English Shires are one of the world's oldest breeds dating back at least 1,000 years. These are the horses that carried knights into battle in the Middle Ages, and they have been used for farm work of all kinds, including pulling heavily loaded delivery wagons in both town and the countryside. A Shire may easily weigh 1,800 to 2,000 pounds, and a characteristic feature of the breed is abundant hair (called feathers) below the knees and hocks. Gathered in front of a barn owned by the Church of the Firstborn, these gentle draft horses are at home in the shire, or county, of Robertson. *Submitted and photographed by Connie Hardin.*

In Bedford County, the home of the National Walking Horse Celebration in Shelbyville, horses can be purchased at several locations, including Martin Farms. *Submitted and photographed by Jeremiah Richards.*

Petie Warren raised Tennessee Walking Horses for many years in the barn that his father built and was still riding in shows into his 80s. Mrs. Curtis Warren is the owner of the Humphreys County barn, part of which dates to the late 1800s, that has a long history of horse and mule trading. *Submitted and photographed by Jean Owens.*

Before becoming the property of the State of Tennessee, the complex that is now Ellington Agricultural Center (Davidson County) was the farm of Rogers Clark Caldwell and known for its livestock operations in the 1920s. The barn that now houses the Oscar Farris Agricultural Museum is believed to have been the thoroughbred horse barn. *Courtesy of the Tennessee Department of Agriculture.*

Below, horses come from all over the United States to be trained for police work at Ellington Agricultural Center, headquarters of the Tennessee Department of Agriculture in Nashville. The mounted patrol facility includes stables for 44 horses, a training arena, and grazing areas. Sergeant J. D. Harber of the Metropolitan Nashville Police Department is in conference with Bear. *Submitted and photographed by Larry Merritt.*

This barn on Lee's Creek Road in Lincoln County has provided shelter to generations of horses. *Submitted and photographed by Kim Carlisle.*

In the Whiteville community of Fayette County, this late 19th-century barn owned by Randy Pepper has an ample hayloft, eight horse stalls, and a tack room. *Submitted and photographed by Ashley Pepper.*

Harlinsdale Farms (Williamson County) has long been associated with the Tennessee Walking Horse industry. The City of Franklin recently purchased the farm where horses continue to graze in the equine park. *Photographed by Caneta S. Hankins.*

This spacious and beautifully landscaped barn in Bedford County speaks to the comforts afforded Tennessee Walking Horses. *Submitted and photographed by Nancy Adgent.*

MULES

Mules are recognized first by their long ears. Sometimes mistaken by the uninitiated as horses, these animals can be contrary and so willful that the phrase "stubborn as a mule" became a well-known description for people. Mules are hard-working animals, respond to short commands such as "gee" for a right turn and "haw" for a left turn, and are often family pets as well. Long an essential part of Southern American agriculture, mules were mentioned by Thomas Jefferson in the farm book he kept from 1774 until 1826. In Tennessee, mules are the most enduring of the working farm animals. Many a family made their crops of cotton, tobacco, and corn as well as a vegetable garden with just one mule, and it often doubled for transportation. A good mule was, and still can be, expensive. In 1870, the average price for a mule was $106, compared to $85 for a horse. In today's market, mule prices vary from about $1,000 to $3,500 for those already trained for riding or pulling farm machinery.

Breeders and owners of mules praise them as being smart, hardy and long-lived, strong, sure-footed, and able to withstand the heat of Southern summers. Mules pulled wagonloads of cotton, corn, hay, and logs; helped clear fields of stumps and rocks; broke new ground; and powered molasses mills, mowers, and threshers in the days before motorized machinery. The mule also contributed to the basic agricultural economy of the state for decades. Breeding mules, which are a hybrid cross between a donkey stallion or jackass and a horse mare, along with training and selling the animals, was big business across the state until the mid-20th century. Some of the largest barns in the Volunteer State were

For 60 years, J. Olin and Dorothy Catlett farmed their Sevier County land acquired from his siblings after his parents' death. In this undated photograph, submitted by his daughter, a young Olin shows a team of mules in front of one of the 19th-century barns. *Submitted by Sandy Bartlett. Photographer unknown.*

Above, in 1890, Pink Gray of Lincoln County laid a stone foundation on which he erected a barn that is 52 feet wide by 120 feet long with a loft for hay and grain. Bill Allen, the current owner, writes that "Mr. Gray would keep 50 to 60 pairs of mules in this barn to work the crops." *Submitted and photographed by Bill Allen.*

built especially for mules. In 2004, the Tennessee Department of Agriculture reported that there were nearly 21,000 donkeys and mules in Tennessee. You may occasionally see a single mule plowing a vegetable garden or a team pulling a wagon, but more often they appear in harness at county and regional fairs, in parades and special events such as Columbia's long-running annual Mule Day, and at farms that welcome the public.

Plez Akins, picking up part of a mule harness, was a tenant farmer for Jessie DeLay, who built this barn in 1920. The barn, now owned by Sam DeLay, is in Rutherford County. *Submitted and photographed by Joe DeLay.*

Carrying on traditional work at the Minton farm in Claiborne County, a team pulls a wagonload of hay from the barn built in the 1920s by Creed Minton and J. W. Cheek. *Submitted by Lewis Minton. Photographed by Ruth Minton.*

Thomas Steagall is ready to work his team of mules on his Rockvale farm in Rutherford County. The early 20th-century granary built by Ben Taylor stored corn for the mules and other livestock. *Submitted and photographed by Sandra S. James.*

William Walker began building this mule barn in Hickman County in 1860 but had to wait until the Civil War ended to complete it. Fred D. Baker owns the barn that retains its stalls, loft, and hayracks. *Submitted and photographed by Donna Baker.*

Diemer and Bob Smartt and 12 other men are responsible for constructing this 1940 barn near Smartt Station (Warren County). They used only hand tools, timber cut from the lot, and 1,000 pounds of nails. The barn is 88 feet long, 80 feet wide, and 40 feet high with a loft that contains approximately 120,000 cubic feet, providing maximum space for loose hay. The barn could accommodate 16 mules. The three generations shown here are Martha, Will, and owner George Smartt. *Submitted and photographed by Ben Newman.*

The "Mule Barn" on the County Line Century Farm (founded 1869) was built in 1918 for $300 by Hooper Linton. At the barn with built-in mangers is Lloyd Linton, the son of Hooper, along with Millie the donkey and Sadie the dog. Lloyd and his wife, Frances, enjoy living on the historic property that is located in both Davidson and Williamson counties. *Submitted by Lloyd Linton. Photographed by Mary Linton Little.*

Porter Elder built this barn in 1925 and always kept six to eight mules in it to work his Pickett County farm. When you see a Tennessee agricultural license plate, this is the barn that inspired the design. It was created by Elder's granddaughter, Cynthia Kent, who works at the Tennessee Department of Agriculture. *Submitted and photographed by Maria Huddleston.*

Located off Highway 100 in Chester County is a gambrel-roof stock barn built in 1934 by H. G. McGee and John Harris. Douglas Jones, owner with LaVon Jones, explains that the barn, one if not the largest in West Tennessee when built, was "large enough to turn a wagon with six mules around." *Submitted by Douglas C. Jones. Photographed by LaVon Lemons Jones.*

Visitors to Epperson Springs in Macon County were met at the train station and conveyed to the hotel complex in wagons pulled by teams of mules or horses that were stabled in this barn. *Submitted and photographed by Judy Binion.*

Dairy Cattle

Settlers moving into what is now Tennessee by routes such as the Cumberland Gap often carefully led the all-important milk cow as one of their prized possessions. If they did not have one, they soon bought one or more when they purchased a piece of land and began to farm. It was not until after the Civil War, however, that the dairy industry truly flourished in Tennessee. In the decade following that conflict, Holstein and Jersey cattle were imported, and herds were established across the state. The Holstein, or Holstein Fresian, originated in the Netherlands, and Thomas M. Spofford of Giles County is credited with first importing this black-and-white breed. Holsteins are known for producing large amounts of milk with a lower content of fat than most other milk cows. Jerseys, from

In her 92 years, Bertha E. Cothran milked many cows on the family's Tipton County farm. Daughter Jean remembers that milking became one of her chores by the time she was 8 or 9. *Submitted by Jean Jordan. Photographer unknown.*

the Isle of Jersey in the English Channel, were brought into the state just after the Civil War, and the Tennessee Jersey Cattle Club was established in 1872. Tennessee Agricultural Commissioner O. E. Van Cleave promoted Jersey cattle at the State Production Show in Lewisburg in 1930, and from 1936 to 1940, Marshall County was the nation's leading Jersey county.

The invention and widespread use of the cream separator in the early years of the 20th century made the milk of Jersey cows, known for its high fat content, more profitable for dairy farmers. Throughout most of the 20th century, the dairy industry was an integral part of Tennessee's overall

These current residents relax between milkings in the shade of the Weakley County barn owned by Gregory Naney. *Submitted and photographed by Trenia Reagan.*

"The Milk Barn" was constructed by Raymond Ricker in 1954 as part of the farm improvement program of the University of Tennessee Agricultural Extension Service, the Pet Milk Company, and DeLaval Electric Milking Company. According to their granddaughter, Melanie Justis, the barn was built to the design of "a 'state-of-the-art', V-type, milking barn" and continued to be used for milking by Raymond until his death in 1991 and until 1994 by his wife, Louise, who died in 2005. John and Phyllis Ricker own the milk barn that was used twice a day for 40 years. *Submitted and photographed by Melanie Ricker Justis.*

economy and a major part of the life and livelihood of many farm families. In 1948, the state counted 592,000 head of dairy cows with a production of more than 2 billion pounds. Many local creameries operated in Tennessee, purchasing milk from area farmers and processing milk, cheese, cream, and ice cream for in-state sale. Others sold quantities to larger dairy processing concerns in other states. The decline in the dairy industry in Tennessee in the past two decades is due to development and loss of farmlands, lower prices for milk and dairy products, the lack of workers to help with daily milking and related chores, and the decline in the number of farming families who can and choose to try to make a living from a job that requires a commitment of time and effort each and every single day. In 2006, only 70,000 dairy cows were reported in the state, and the production was between 1 million and 2 million pounds. East Tennessee was the highest producer in the dairy industry with 36,000 head. Middle Tennessee, traditionally the largest producer, reported only 18,500 milk cows. Dairy barns for herds of all sizes still claim a good part of the agrarian landscape, bearing witness to what was once a vital part of the state's agricultural economy and farming lifestyle.

From 1940 to 1945, Mr. and Mrs. W. H. Roberts owned the Greenvale Dairy near Jackson (Madison County). Because work required Mr. Roberts to travel, it was Mrs. Roberts who would walk about a mile around 3 a.m. and again in the afternoon to milk and feed the cows. Their grandson, Glenn Cooper, is restoring the dairy barn and keeps these bottle caps as a reminder of the family's dairy enterprise. *Submitted and photographed by Glenn Cooper.*

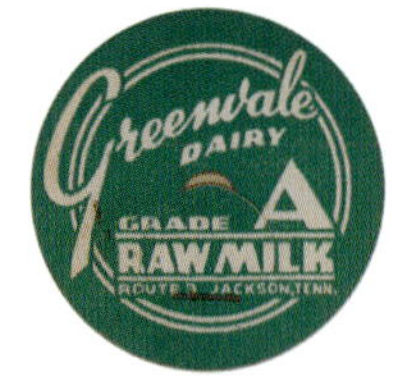

Cows begin to gather at feeding time around this Union County barn. *Submitted and photographed by R. F. Ellis.*

The dairy barn, flanked by matching silos, is at the heart of this busy Fayette County farm. *Submitted and photographed by Van Jolly.*

A Holstein herd grazes in front of a Marshall County dairy barn in a county that once led the nation in Jersey milk production. *Submitted and photographed by Kathy Boice.*

"Selling milk has meant a better standard of living for our family, better furnishings for our home, and more good equipment on the farm. Dairying helps to keep good tenants, too, and the milk cows have improved our land greatly."

From George Birdwell Sr., as quoted in the 1953 "Pet Dairy Chats" newsletter of the Pet Milk Company and provided by the Birdwell family, Still Hollow Century Farm, Greene County.

Above, J. S. Campbell built the barn in the late 1940s, and Tom, Roy, David, and Joe, who is responsible for the sign, are the owners of the Campbell Brothers' operation in Crockett County. *Submitted by Joe Campbell. Photographed by Scott Lovelace.*

River Valley Farm, bordered by the Barren Fork River as it runs through Warren County, has been in the Barnes family for five generations. The entire family was involved in dairying and also raised and showed registered Holstein and Guernsey cows at county fairs across Tennessee. The barn, built by Vickie Bouldin's grandfather, Doyle Barnes, and others, has several stables where newborn calves and their mothers waited to be added to the herd. *Submitted and photographed by Vickie Barnes Bouldin.*

A little snow could not deter Mr. and Mrs. Emmett Jenkins of Nolensville from heading to their Williamson County barn at milking time. *Submitted and photographed by Betty J. Hughes.*

The Cleburne Jersey Farm in Spring Hill is one of the most historic dairy farms in the nation. In 1964, the Tennessee Jersey Cattle Club, established in 1872 and the oldest such organization in the nation, declared that the Cleburne Jersey herd was the oldest in the United States. The National Register nomination for the Maury County farmstead notes that the original stone dairy and separator house, above, dating from the 1870s and constructed soon after the family emigrated from Scotland, may well be the oldest such complex in Tennessee. *Courtesy of the MTSU Center for Historic Preservation.*

Today, the milking and feeding at the Cleburne Farm in Maury County takes place in 20th-century barns. Owner Pat Campbell continues the tradition of one of the premier dairy farms of Tennessee. *Submitted and photographed by Lisa Tidwell.*

The interior of the 1940s barn owned by Severt Allen in Hawkins County gives a glimpse of the equipment needed to run a milking operation. *Submitted and photographed by Tammy Murrell.*

The instructions, above, are clear to anyone who might have shown up at the hours of 4 a.m. or 4 p.m., when milking was done at this Grade A dairy in Wilson County. B. W. and Karleen Eakes are the owners of this Wilson County barn, at right, built 95 years ago by Birdie Castleman and John E. Wright. *Submitted and photographed by Karen Thompson.*

Mr. and Mrs. Charles Meek on their Williamson County farm. The Meeks operated a dairy here until 1983. "A farm is a good place for children," said Mrs. Meek. *Submitted and photographed by Robert Chambers.*

John A. White, who cut the logs and made the wooden shingles for this barn in the 1930s, milks the family cow. The White family of Gibson County still owns the barn. *Submitted by Robert D. White. Photographer unknown.*

Beef Cattle

The breeding, raising, and selling of beef cattle is the No. 1 livestock industry in Tennessee today. While market prices are always in flux and 2007 was a disaster for many farms, cattle continue to be raised across the state on farms large and small and on terrain that is grassy, rocky, hilly, or flat where dairy herds used to graze and row crops were once grown. The Aberdeen-Angus, or Angus, first brought to the United States from Scotland in 1873, is the most popular breed raised in Tennessee. Angus beef is highly desirable because of its characteristic marbling of fat and lean, and the breed has led all others nationally since 1963 in this regard. Charolais, originating in France, is a leaner beef that appeals to the more health-conscious consumer. Herefords also have been successful in Tennessee and are favored by many cattle growers. The Tennessee Cattlemen's Association, founded in 1982, protects and promotes the beef

Bovine residents of River Circle Farm in Trousdale County graze on Martha Dixon's property, where a 1920 barn is among the several outbuildings that support the farming and cattle operation. *Submitted and photographed by Kristi Shirah.*

industry and provides information and education for cattle growers across the state. In 2006, Tennessee ranked ninth in the nation's beef cow inventory, and sales brought $514 million, which accounted for 20 percent of farm cash receipts received by Tennessee cattle farmers. Greene, Lincoln, Giles, Maury, and Bedford are the top cattle-producing counties, and more than half of the state's farms raise cattle.

Long before this barn was built in 1910, horses and cows were hidden among the bluffs of this Cannon County farm during the Civil War. After four generations, the Thomas family continues to raise cattle on the 130 acres. *Submitted by Gary L. Thomas. Photographed by Judy Thomas.*

On Milton Road in Rutherford County, cattle watch passersby from their grazing area around the red barn. *Submitted and photographed by Lori Manns.*

This barn, used for bottle-fed calves, contains a corncrib and shuck pen as well as a large hayloft and cattle chute. Mr. and Mrs. Liston Hitchcock of Bone Cave in Van Buren County make good use of their barn built by Murphy Hitchcock and several other carpenters in 1932. *Submitted and photographed by Chelsea Roberts.*

Built in 1885 in Greene County, the Robert L. King barn, above, once stabled work horses and is now used for working cattle. *Submitted by Carol Lynn Coffman. Photographed by Danise Nelson.*

The cattle and hay barn, left, dominates the landscape on the Wilson County property owned by the photographer. *Submitted and photographed by Diane Tatum.*

Dwight "Dee" Cooper shows off his prize Angus bull in front of the Cooper family barn in Hawkins County. Stacie Hill is grateful to have this photograph of her great-grandfather. *Submitted by Stacie Hill. Photographer unknown.*

Some 70 years old, this Williamson County barn still welcomes cattle at feeding time. Hardie and J. T. Williams built the barn, which still stands at Rolling Acres Farm in Nolensville. *Submitted by Kaye Burns and Peggy Taylor. Photographed by Jenn Duncan.*

Shane Campbell is the fourth generation to own the Claiborne County farm. Built in 1919 for Nelson and Martha Campbell, the barn was constructed of chestnut, white oak, and pine. Used for tobacco and storage of hay and corn for the farm's livestock, cattle are run through for vaccinations. *Submitted by Shane Campbell. Photographed by Lori Campbell.*

The original barn is about 120 years old, and the enclosed spring now waters the registered Pinzgauer herd of Mary Ann Meredith of Robertson County. The name comes from the "Pinzgau" region of Salzburg, Austria. Pinzgauers have a mild disposition, are good producers of milk, and the meat is of fine quality with good marbling. *Submitted and photographed by Susan Morley.*

The Haywood County livestock barn owned by Wayne and Patsy Tritt was built in the late 1920s or early 1930s by Henry Waddell. *Submitted and photographed by Wayne Tritt.*

Norris Bringle and his sons built this massive Tipton County cattle and hay barn in 1948. It is 70 by 70 feet, has two sheds for feeding cattle, five stalls for mules, six storage bins, a large corncrib, and two separate sections in the loft so hay can be pushed through to the feeding troughs. The loft can hold 7,500 rectangular bales. The Bringle family continues to raise cattle and make good use of the barn. *Submitted and photographed by Becky Bringle.*

Cattle stay close to the hay barn on the Westbrook Century Farm (established in 1887) in Williamson County. Mary Anne Warren is the current owner of the family farm and this 80-year-old barn. *Submitted and photographed by Carolyn Simpson.*

Other Livestock

Sheep

In the years just prior to the Civil War, Tennessee was the 10th largest producer of sheep in the nation. Though the number of sheep raised never compared to that of hogs, goats, and cattle, they appeared with the early settlers and remain a part of Tennessee's agricultural tradition, both for wool and meat. Lamb and mutton, not as popular as beef and pork, remain an inconsistent market. With the introduction of nylon in the 1930s and other synthetic materials since that time, wool ceased to be a cash commodity for decades. A return to popularity of natural fibers including linen, cotton, and wool has helped some farmers maintain small flocks or return to raising sheep. Cottage industries that produce high-quality woven goods also enable some farms to keep sheep either for demonstration and education or to supply weavers and small textile companies. The proper shearing of sheep is difficult and requires skill and knowledge, and few laborers are available to assist with that task. For these reasons, flocks across Tennessee — 25,000 head were counted in 2007 — tend to be small. Beyond the limited marketing options, sheep are also more likely to be victims of predators and require regular attention. In the absence of full-time shepherds, the owner, usually with the assistance of a herding dog or a protective donkey, must move the flock to a pasture each morning and then return them to the shelter of a barn for the night. Lambing time requires a special watchfulness on the part of sheep farmers. The sight of sheep and lambs grazing on a hillside or being herded to or from their barn continues to be a pleasure for the passerby and, apparently, worth the effort for a few farmers as well.

These recently sheared sheep follow their leader and protector to the barn on Clyde Hall's farm. Hall trains border collies, and national trials have been held at his farm in Murfreesboro, where sheep play their traditional and important part in the competition. *Submitted and photographed by Renee Lowery.*

Used for decades for sheep and pigs, part of this log barn in Jefferson County may be at least 200 years old, according to owner Louis Vaccari. *Submitted and photographed by Louis J. Vaccari.*

"Feeding time" finds Willard Pearsall of Grundy County in the yard of a barn built by J. J. Walker in 1889. Mrs. Goodson, Pearsall's daughter, remembers that the barn had "a tack room, corncribs, hayloft, lots of stables, and a big center hallway where the wagon stayed and the cows were fed while being milked." *Submitted by Katharine P. Goodson. Photographed by E. B. Blatz.*

Swine

Though the United States ranks second in the world as a producer, consumer, exporter, and importer of pork and pork products, Tennessee no longer makes a significant contribution to this market. But at the end of the 1850s, as the Civil War approached, Tennessee ranked fourth nationally in swine production. Only about 2 percent of the state's farms now raise pigs, though 65 percent of farms included hogs in 1950. As late as 1988, more than 1 million hogs were reported in Tennessee, but the count declines each year with only 356,000 head in 2004. Historically, large and small farms alike raised hogs. A small farm might have a single sow that produced a litter each year. From this litter, a few might be sold for cash while the remaining pigs were fattened and slaughtered during the first truly cold spell of the year, usually after the end of November. Smoked hams, sausage, and bacon became a mainstay of meals year-round. Tenderloin and spare ribs are favored parts. Cracklings, left over after lard is rendered, are the original pork rind snacks and are also used in a regional specialty called "cracklin' bread." Lard was the cook's primary shortening and used for frying and baking before the days of margarine, vegetable shortening, and oil. The enjoyment of chitterlings and souse is an acquired taste, while hog jowl and other pieces may be used to season beans and other vegetables. Pig's feet, either boiled or pickled, are considered a delicacy by many. Smoked ears and hooves are prime dog treats. While not entirely true, the old saying that "everything was used but the squeal" is close enough.

Most farms, even well into the 20th century, had smokehouses to cure and store hams, shoulders, sausage, and bacon. Smokehouses, including this one built in the late 1890s by Lewis Stinson of Macon County, are found on many farms. Though they still retain the smell of salt, smoke, and meat, they are rarely used for their original purpose today. *Submitted and photographed by Chelsea Compton.*

A 19th-century log building, above, in Petersburg (Marshall County) housed piglets and hogs for so many years that their constant "rooting revealed the beautiful rock formations (with lots of fossils) and formed our back yard rock garden," explain current owners Sam and Betsy Craig. *Submitted and photographed by Betsy Craig.*

Left, surveying his Berkshires in the 1920s, is Madison Lafayette Southall of Tipton County. The barn, built by Southall in 1910, has eight mule stalls and a generous hay loft. Side sheds and feed troughs for cattle were added in the mid-1950s. Robert Southall Nix is the owner of the barn today. *Submitted by Milton Southall. Photographer unknown.*

This herd of goats knows this barn, built in 1927 in Buchanan (Henry County) by Charlie Sweatt, father of the current owner, is a good shelter. *Submitted and photographed by Charles A. Sweatt.*

Sandy and Dale Sanders of Marshall County believe that Buford Jones built their barn in the 1920s. They use it to house their herd of ADGA purebred Nubian goats. *Submitted and photographed by Sandy Sanders.*

Goats

Raising goats is one area of livestock production that has significantly increased during the past decade. In 1997, the Department of Agriculture reported only 50,000 head across the state. In 2002, the first state-graded goat sale was held, and in that same year 107,000 meat goats were reported, making Tennessee the second-largest producer in the nation. Goats, like sheep, have long been a part of the agricultural landscape, keeping weeds and grass under control and supplying an alternative to dairy products. Some breeds are grown especially for their milk while others supply hair that is woven into high-quality sweaters and other items. Goats are fairly resistant to diseases but require specific management practices and need protection from predators. Guard dogs, such as the Great Pyrenees, and donkeys are the usual keepers of goat herds.

Until the arrival of increased numbers of immigrants into Tennessee from countries that traditionally grow and eat goat meat, producing goats for food was not a lucrative market. Today, however, farmers are taking advantage of the expanding consumer market that includes people who have always enjoyed goat meat and also those who are health-conscious. Chevon, the more formal name for goat meat, is lean, low in calories and cholesterol, and high in protein. The Tennessee Goat Producers Association was formed in 2002 to provide a dependable market for Tennessee goat farmers and their products. The association's Web site advises that goats are friendly, even "charming" animals and can thrive under conditions that would not support other livestock. Goats are a favorite animal for petting zoos and for farms that provide educational tours and activities for school groups and the general public.

When Tolbert Conley built this barn in 1880, would he have imagined that animals unknown to Tennessee and rarely seen in North America would find shelter in it? The safari park at Hillcrest Farm in Crockett County makes good use of this livestock barn. A kangaroo looks on as owner Claude Conley has a discussion with a camel. In the past several years, a growing number of Tennessee farmers have imported and are raising exotic animals, including emus, alpacas, llamas, and ostriches, for a variety of purposes. *Submitted and photographed by Tina Marie Brookes.*

Poultry, including geese, ducks, turkeys, and chickens, were once found on almost every farm. Raised for eggs, meat, feathers, and down, chickens and geese, especially, let people know when strangers and potential predators appeared in the farmyard. Chicken coops, hen houses, and chicken houses were either separate outbuildings or spaces within barns that could be secured each night to protect valuable but vulnerable fowl. Watching the parade of geese in front of a pre-Civil War barn in Perry County are owner Willidean Kilpatrick, son Jamie, and granddaughters Haley and Hana. *Submitted and photographed by Elizabeth DePriest.*

When his grandson, Rusty, decided to raise goats for his Future Farmers of America (FFA) project a few years ago, Russell Byrum constructed this goat barn. *Submitted and photographed by Russell Byrum.*

Cats and Dogs

For centuries, cats and dogs have played important and varied roles in farm life. Traditionally, cats are kept to control vermin from infesting and eating stores of grain and feed. Dogs move livestock from barn to pasture, provide protection from predators, and act as the farm's early warning system. Over the years, breeds and types of livestock and crops change, but the feline and canine workers, companions, and playmates are a constant on farms and especially around barns.

At the Detwiler farm in Shady Valley (Johnson County), left, this lone patrol keeps watch over everything, including the barn built by Tom Hubbard in 1990. *Submitted and photographed by Lori Detwiler.*

The guardian of this Grainger County barn is on "high alert." *Submitted and photographed by Rick Bunge.*

A cat heads to work on this Lewis County farm. The barn was built by A. J. Barber in 1947-1948 as a small dairy that operated for more than 20 years. The building suffered some storm damage several years ago and is used mostly as a loafing place for the cattle that Gene Barber, A. J.'s son, raises. *Submitted and photographed by Linda Parker.*

Barn cats in the Turkey Creek area of Clay County pause to see what is happening outside their barn. *Submitted and photographed by Heather Burnette.*

"George is always ready to go," writes Chuck Billingsley of his faithful blue heeler. Charlie Rogers built this Putnam County barn in 1986 to replace one that burned after being struck by lightning. *Submitted and photographed by Chuck Billingsley.*

Though this Overton County barn was built in 1960 for the family's Tennessee Walking Horses, it is now home to a kennel where Mountain Feist dogs are raised. Alvin Huddleston advises that these dogs are bred to hunt squirrels primarily but also will take on rabbits, birds, and other small game. "Mo" is anxious to be out of the barn and about his business. *Submitted and photographed by Maria Huddleston.*

This elusive but very much "on-guard" barn cat represents the many generations of felines who earn their keep by protecting corn and other feed from rats and mice. This one lives in Mandy Duncan's barn in Gravely Valley (Hawkins County). *Submitted and photographed by Stacie Hill.*

"This English Setter was my all-time favorite quail dog," writes Charles Owen of Humphreys County. The photograph was made about 1980; the barn burned in 1985. *Submitted and photographed by Charles Owen.*

Where there is milk, there are likely to be cats nearby. This cat has staked out territory in the 1817 barn that is part of the Charit Creek Lodge in Sevier County. *Submitted and photographed by Heidi Fincham.*

Field Crops

For more than 200 years, corn, tobacco, hay, and wheat, along with cotton, have topped the state's field crop production list. In February of 2008, the National Agricultural Statistics Service reported that cropland accounted for approximately 60 percent of the total farmland in Tennessee. In national rankings for 2007, the state was fifth in the production of snap beans and tomatoes while tobacco was sixth and cotton and hay placed ninth. Corn and soybeans also continue to be among the most popular and valuable crops. The following pages and images focus primarily on barns associated with the state's traditional and familiar field crops.

Corn was the first crop grown by settlers who came into what would become Tennessee and was also one of the early crops to be marketed for cash. In the 1850s, the state's corn production was sixth in the nation. Corncribs are among the oldest of stand-alone outbuildings and are also common spaces within barns of all sizes. Corn continues to be grown across the state in small vegetable gardens and in large fields. With advanced processing methods and economic incentives for producing biofuels — particularly ethanol, which uses only the starch of the corn kernel and still leaves the remainder for other uses — corn production will likely continue its rise in Tennessee in the future.

The southern counties of Middle and West Tennessee are part of the South's traditional cotton belt and, as natural fibers have become popular again, this primary cash crop continues to be grown. As far as barns are concerned, cotton was generally a crop that went on to be ginned as soon as possible. Wagons loaded with bales may have sheltered in barns for a time, but only the amount needed for spinning into thread or for other purposes was stored. Equipment, scales, and sacks, used until the 1960s when mechanical cotton-pickers appeared in the state, were stored in nearby barns.

Soybeans are not stored in barns, and the greenhouse/nursery industry, which is the second-largest non-livestock cash producer, requires specialized facilities rather than traditional outbuildings. Irish and sweet potatoes often had their own buildings where these common vegetables were harvested for sale and where the family's share was stored in dry spaces that stayed above freezing. Orchard fruits and berries are sometimes stored in barns for short periods of time for wholesale and retail sales or before processing. Apples, blueberries, and Tennessee's perennial favorite, strawberries, can be found across the state, often at pick-your-own farms.

Built in 1956 by Albert Warren and Wesley Williams for their father and father-in-law, Judge W. T. Warren, the barn is a center of activity on this farm that for years supplied a variety of vegetables for the Farmers' Market in Nashville. The straight rows of this large garden were planted and are being cultivated by Albert Warren, the current owner of the farm near the Warren and Cannon County line. *Submitted by Albert Warren. Photographed by Judy Warren.*

In Crockett County, 70,000 acres were planted in cotton in 2006. The barn is surrounded by the crop that continues to be an important part of West Tennessee's agricultural economy. *Submitted and photographed by Tina Marie Brookes.*

Rows of corn are a staple in gardens like this one in Bledsoe County owned by Mauvine Brown. The general livestock and crop barn was built by Clinton Brown in 1949. Wheat, though not grown in the quantities of the 19th century when Tennessee was a leading producer in the nation, brought more than $42 million to the state's farmers in 2006. Barns and granaries, built specifically for threshing and storing wheat, are among the finest and most enduring of outbuildings. *Submitted and photographed by Margaret Thurman.*

The stand-alone corncrib on Danny Vance's farm in upper East Tennessee (Carter County) was built about 50 years ago by his father. *Submitted and photographed by Jan Rushing.*

This corncrib, left, has been in constant use since owner Gilbert Justice and his father, John S. Justice, built it in Cheatham County just after World War II. *Submitted and photographed by Amy Harvison.*

Watt Nolan built this Sumner County barn, right, in 1910. The corn and grain bins are constructed with wooden pegs. *Submitted and photographed by Frank Coley.*

This massive granary was built by Henry Haney in the early part of the 20th century. Located in Trousdale County, it is owned by H. B. Cunningham today. *Submitted and photographed by Sabra Todd.*

Three generations of the Teeter family, left, in the hallway of their Montgomery County granary, which was built by James Johnson in 1880. *Submitted by Laurence Teeter Jr. Photographed by Mike Jones.*

Built by slaves on the Fermanaugh-Ross Century Farm in Greene County, this wheat-threshing barn dates to 1825. Carol Lynn Coffman owns the working farm that has been in her family for nearly 200 years. *Submitted by Carol Lynn Coffman. Photographed by Danise Nelson.*

The "Strawberry Barn" and its totem pole are unique to Wayne County. *Submitted and photographed by Michael T. Gavin.*

Now 80 years old, the granary for Corbin Dairy in Robertson County is still in use. *Submitted and photographed by Amy Hooberry.*

Wheat threshing at W. T. Carathers' farm in the Knob Creek community of Washington County. The large barn was built from oak timber and pegged together. *Courtesy of the MTSU Center for Historic Preservation.*

This barn, below, built in the 1940s by the Miller family and owned by Bob Schmidt, is an attraction for customers who come each season to buy strawberries and jam at the Maple Lane Farms in Blount County. *Submitted and photographed by Jeff Spurling.*

Though built in 1907, this type of corncrib was constructed and used for more than 100 years. Examples like this one on Muriel Bradley's Century Farm in Hardin County can still be found in many counties. *Submitted by Debra Ann Bradley. Photographed by John F. Bradley.*

Hay

Across the United States, hay has traditionally followed corn as the most important crop. Only in the last half of the 20th century did soybeans move to second place in Tennessee, with hay following third. In 2006, hay was ranked 10th in the leading commodities for cash receipts for the state's farmers. In that year, 210,000 acres were harvested with a yield of 2.3 tons per acre. Alfalfa, a common and productive strain of hay of Spanish origin, is often used as a rotation crop to replenish nitrogen in the soil. Alfalfa production is reported separately from other varieties, which include fescue, lespedeza, clover, timothy, and orchard grasses as well as mixed hays.

Barns used to store hay range from large livestock barns with haylofts occupying the entire second floor to small sheds and covered open-air facilities. Bulky but lightweight, hay is kept near animals yet out of the way in the loft where it can be dropped down into mangers or onto wagons or truck beds. Until automated balers appeared in the late 19th century, hay was stored loose, necessitating the large spaces seen in 19th-century hay and livestock barns. Rectangular bales, commonly referred to as square bales, allowed for more compact storage and for easier transport of hay to herds in outlying fields.

The most common type of Tennessee barn, referred to as the central aisle (see Part I), often features a bonnet or extended roof from the loft. It was designed so that a system of pulleys could move loose hay from wagons along the line of the ridge of the barn to a spot where it could be dropped.

Later, conveyor belts moved rectangular bales up to the opening of the loft. Underneath the bonnet, hay could be moved from the mow and pushed out to vehicles waiting below. In recent years, round bales that can be moved by one person using a hydraulic lift have become more popular. Many farmers, however, continue to cut and bale in the familiar rectangular block. In August of 2007, a year that brought a generally disastrous growing season and harvest, many farmers had to resort to buying hay to feed the cattle they did not sell. Of the 19 farmers listed on the Tennessee Farm Bureau's hay directory as having surplus hay for sale, 18 offered rectangular bales, no doubt because they could be loaded easily and transported wherever needed. While a few farmers reported their hay was in both field and barn, most indicated their bales were in barns. Because hay is the most common crop stored in barns, these buildings are found in every county and range in condition and age from the early 19th century to the present.

Granddaughter Isabella Adalisha Gray, daughter Martha Shelton Gray and Willie Shelton with the barn Willie's father-in-law, John Oliver Petty, built in 1948. The McNairy County barn was built from timber harvested on the property. The barn has a hay carrier and a corn crib. *Submitted by Martha Gray. Photographed by Brent Gray.*

Hay fields and barns, left, are a familiar sight in Roane County and across Tennessee. *Submitted and photographed by Michael Irvin.*

Above, the Berry family of Union County has used this barn for three generations to hang tobacco and store hay. *Submitted and photographed by Brittany Berry.*

The hay and livestock barn on the John C. Kemmer Century Farm, established in 1879 in the historic farming community of Grassy Cove in Cumberland County, was built in the 1920s. John C. Kemmer IV and his family continue to raise hay, cattle, and sheep. *Submitted and photographed by Debbie Hill.*

This 2002 hay storage facility in Rutherford County, constructed by Donald Blankenship, uses 300-pound concrete blocks to anchor the curved steel rafters that support the fabric cover. The design is quite efficient for storing hay and reduces the labor needed to manage the round bales to two people. Helen Blankenship is the owner. *Submitted and photographed by Roger Pruitt.*

The conveyor is in place to move hay bales into the loft of Steve Kennedy's barn in Fairview (Williamson County). The sunburst door is an unusual and original feature incorporated by builder Frank Carter in 1940. *Submitted and photographed by Kathy Jones.*

Loose hay is moved into the loft by a series of ropes and pulleys at this Scott County barn. Built in 1949 by Arthur Elliott from white oak sawn and milled on the property, it is owned by Sarah Jones and Milton Smith; he works the farm with Belgian horses that are housed and fed in the classic livestock/hay barn. *Submitted by Rachel Storey. Photographed by Donna Storey.*

Tobacco

Tobacco is one of the older crops cultivated in North America and in the Southeastern United States. We know from the elaborately decorated pipes of the Cherokees that tobacco was widely used in their ceremonies and rituals. Tobacco was grown commercially in the colonies as early as 1612 and came into East Tennessee with the settlers of that region. Prior to the Civil War, tobacco was second only to cotton as a cash crop for Tennessee farmers. Unlike cotton, large acreage was not necessary for a good yield, so while there were some large farms and plantations that produced tobacco, it is more often associated with small- and medium-sized farming operations. Because a tobacco crop requires hard physical labor, and is nearly a year-round responsibility, an adequate work force is necessary. Large and extended families worked together, along with neighbors and hired hands, to strip and hang tobacco. Plantations like Wessyngton in Robertson County, which was the largest tobacco producer in the state and perhaps the country prior to the Civil War, relied on slave labor. Two main types of tobacco, dark-fired and burley, are grown in the Volunteer State, and each requires a barn with specific characteristics.

In the northern counties of Middle and West Tennessee as well as some counties in the upper eastern part of the state, the darker varieties of tobacco are grown and fire-cured. The leafy tobacco stalks are hung in vertical tiers (or rooms) inside a tall barn with characteristically strong vertical lines. Small piles of rotten wood or sawdust are lit to produce plenty of smoke but little fire. Smoke exits via openings in the eaves. Care must be taken to produce the exact conditions to properly "fire" the tobacco because too little heat will cause the leaves to mold, and too much heat could destroy the barn.

By 1860, the burley tobacco had been introduced and was becoming a more popular variety. Burley is easier to cultivate and is cured by air that circulates through the tiers of hanging tobacco leaves. Virtually any structure with a roof is suitable for drying the crop in this manner, so many different barn types have been employed for this purpose. The most important requirement is continuous air movement. Ventilating devices may be placed in a series at the peak of the roof, along the side of the ridge in a continuous vent line, at gables and eaves, and in the walls to facilitate the curing process.

In this vintage photograph taken in the 1920s or 1930s, men pause from cutting tobacco that will be cured in the barn behind them. The barn has been on the Williams farm in Smith County for several generations. Lester Williams, second from right, is Carolyn Chaffin's grandfather. Her father, James, continued to raise tobacco until his death in 1999; her mother, Frances Williams, owns the farm. *Submitted by Carolyn Chaffin. Photographer unknown.*

A Macon County barn seems to be watching over the new tobacco plants as they begin to grow. *Submitted and photographed by Karree Phillips.*

Phil Morris Baggett documented the restoration of a dark-fired tobacco barn that was built in the 1880s by his grandfather, Henry Warren Baggett. The barn is located in Montgomery County in the "Black Patch" region that is noted historically for dark-fired tobacco production. The "tier poles" that hold the tobacco sticks are "small unplaned hardwood trees, many of which still have the bark attached after all these years." *Submitted by Phil Morris Baggett. Photographed by Steve Westfield.*

Burley is favored by cigarette manufacturers while dark leaves are used for chewing tobacco and snuff as well as products that include nicotine in other forms.

Tobacco production in the United States reached its highest level between 1963 and 1975 at 1.75 billion to 2.2 billion pounds. With the health issues associated with the use of tobacco and tobacco products becoming widely known, production has declined dramatically. For example, in 2005, acreage planted in Tennessee was at 22,950, down nearly 25 percent from the previous year, when the "tobacco buyout" was accepted by many tobacco farmers. By 2006, fewer than 20,000 acres were in production. The decline in this traditional cash crop forced farmers to look elsewhere for income and to diversify their commodities. The shift from growing tobacco means that barns are either used for other purposes or else stand idle. Similarly, options for using the large tobacco

Visitors traveling through tobacco-growing counties have sometimes become alarmed to see smoke drifting from a barn. What they are seeing is the curing of dark-fired tobacco leaves in progress. *Submitted and photographed by Lisa Speroni.*

Above, the Bateman family of Houston County is proud of the barn built by Claude R. Bateman more than 80 years ago. *Submitted by Charles D. Bateman. Photographed by Misty Bateman.*

warehouses found in many cities and towns across the state have posed challenges for communities.

Some farmers continue to grow and harvest the crop in traditional ways, but baling tobacco has cut down on the labor required to bring a crop to market. Tobacco is still one of the top 10 cash crops in the United States, which is third after Brazil and China in exports. The Research and Education Center in Greeneville (Greene County), established in 1932, is known internationally for its burley tobacco research and its efforts to improve the net return and products of tobacco growers. Medical research continues to experiment with tobacco — which is, ironically, rich in antioxidants including vitamins A and E and beta-carotene — to determine beneficial uses in vaccines and as a possible cancer-fighting substance. Former tobacco producers either continue to look for crops that will replace the high yield on small acreage or have already made the transition to other commodities. While tobacco growers are few in number today, the impact of this crop on Tennessee's history, economy, culture, and architecture is undeniable.

Dating to 1955, this barn was built to cure burley tobacco on the Van Hooser Century Farm in Cannon County. *Submitted by Sara Van Hooser West. Photographed by C. Van West.*

At right, the Ellis family has raised tobacco on their Robertson County farm for several generations. Their crop of dark-fired tobacco cures in the barn for about a month as smoldering sawdust burns on the dirt floor below. Cap Anderson built the barn for Burgess Gower in 1920. The lumber cost $160. Gower was paid $90 to build it. *Submitted and photographed by Robin Conover.*

Tobacco hangs to cure in a Lincoln County barn. *Submitted and photographed by Bert Hand.*

Bubba Staggs and his father-in-law, Jacky Harris, who built this barn in 1999, continue the long-time Cheatham County agricultural tradition in the Pleasant View community. *Submitted and photographed by Amy Harvison.*

Maurice Elliott looks over the 2006 tobacco crop hanging in a barn built in the 1930s by Richard Hancock. The barn in Big Rock (Stewart County) can store several acres of tobacco; it is owned by Mike Wilson. *Submitted and photographed by Angela Elliott.*

Smoke rises from a Robertson County tobacco barn. *Submitted and photographed by Robin Conover.*

A field of tobacco in front of a Henry County barn. *Submitted and photographed by R. C. Owen.*

When Claude and Ethel Willis bought this Hawkins County farm in 1960, their sons and sons-in-law built the barn. Bud Willis owns the farm and planted this field of tobacco. *Submitted by Bud Willis. Photographed by Laura Willis.*

A Jefferson County tobacco barn and crop. *Submitted and photographed by Rick Bunge.*

The 1915 tobacco barn at Wessyngton Plantation in Robertson County supported the production of the crop into the 20th century. *Courtesy of the MTSU Center for Historic Preservation.*

Above and at left: During the 1991 harvest, Peggy Rory and her mother, Addie Mae, work in the stripping shed attached to a tobacco barn. The shed was built about 60 years ago by McKinley Rory, Peggy's father. The family continued to raise tobacco on their Stewart County farm until the buyout in 2004. *Submitted and photographed by Carol Turrentine.*

Following page, Welbur Ellis checks the dark-fired tobacco crop hanging in a seven-tiered tobacco barn near Cedar Hill in Robertson County. Welbur and his brother, Robert Bell Ellis, have farmed together all their lives. Their great uncle, Burgess Gower, hired Cap Anderson to build this barn in 1920. *Submitted and photographed by Robin Conover.*

Part III

Barns That Deserve Another Look

Introduction

In 2006, agriculture contributed nearly $40 billion to Tennessee's economy, and farmland accounted for approximately 43 percent of the state's total land area. As the 21st century advances, farm acreage in Tennessee decreases each year. Farmhouses, outbuildings, and crop and grazing lands give way to other uses. As these changes take place, we begin to look at the remaining farms and agricultural buildings differently. Are there ways to adapt, restore, and preserve these buildings so they will remain a part of the landscape?

Even as barns are used for traditional agricultural purposes, they often double as billboards and display areas for signs, objects, and artwork. Some are well-known local landmarks that have been enjoyed and photographed for years by residents and visitors alike. Other barns receive a reprieve through restoration that adds years of service to the farming operation. A few owners rehabilitate and remodel for new purposes, finding that barns are very accommodating buildings. It comes as no surprise that the fate of many barns is in jeopardy as the former agrarian landscape is developed in different ways. Some may be included as a part of new land-use plans, but most will disappear. Finally, we take one more look at those barns that no longer exist.

"The individual needs of farmers contributed to the many variations of size and shape that met the demands of the users, but the personality and lovable beckoning of the original concept remained. In scores of hollows and atop hundreds of hills all over the state stand the evidence of this once magnificent asset. Few remain in their original form, and many have been lost to weather, wear, and decay so that those remaining must be treated with respect and dignity so their historical images may be preserved in word, song, and art to ensure their perpetuity."

– Submitted and written by Pelham N. Norris

A barn proudly displaying an advertisement for Ruby Falls. *Submitted and photographed by Robin Conover.*

Signs, Art, and Flags

The steeply-pitched roofs and oversize scale of the sides and ends of barns provide a conspicuous and enduring canvas to display a variety of messages and items. Across the country and certainly in Tennessee, barns have been used to great advantage to advertise tourist attractions, businesses, and products. Flags, artwork, old license plates, signs of many kinds, and various implements and tools find homes on the exteriors of barns. Because of the visibility, longevity, and location of barns, they are popular places that serve as marketing devices, directional landmarks, symbols of achievement, and sights that capture our attention with unexpected art and novelty.

Road Advertisements

In the early decades of the 20th century, the automobile began to have its first significant impact on the United States. People in business, individuals, and families who could afford vehicles and gasoline began to travel improved roads rather than rely on railroad passenger service or fledgling airplane

A piece of folk art itself, this barn's display is a favorite of photographers passing along State Route 133 in Johnson County. Owner Evelyn Miller continues to add to the collection that she and her late husband began some years ago. *Submitted and photographed by Glenna McCracken.*

The Bote Embroidery sign found a home on the Thompson family's barn in White County. *Submitted and photographed by Jane Thompson Bohannon.*

"Signs of Yesterday" is an apt title for a Hawkins County tobacco barn, above. Built in 1983 by Pete Barton and owner Paul Williams, the barn measures 36 by 60 feet and has a stripping shed that is 12 by 60 feet — providing ample exterior space to display vintage signs and license plates. *Submitted and photographed by Stephen Williams.*

service. Gas stations, garages, restaurants, and motels became new features on major and secondary corridors. Sections of the Dixie Highway, built from 1915 to 1927 and extending from Ontario, Canada, to Miami, Florida, came through parts of Tennessee. The eastern part of the road wound from Cumberland Gap through Knoxville, Rockwood, Dayton, and Chattanooga while the western division was routed from Springfield to Nashville, Murfreesboro, Shelbyville, Tullahoma, Winchester, Cowan, and Monteagle into Chattanooga. The Memphis-to-Bristol Highway was designated as State Route 1 in 1915. U.S. highways 70, 64, and 31 are among some of the historic and scenic corridors that still cross Tennessee, allowing the traveler a glimpse of rural landscapes and small towns. It was along these roads that barns were first used to advertise local businesses as well as more distant attractions. Rock City and Ruby Falls are the most famous attractions featured on Tennessee barns, but other businesses and sites also used this inexpensive and effective advertising method.

Rock City was the vision of Frieda Carter, who with her husband, Garnet, developed gardens throughout a natural maze of geological formations atop Lookout Mountain. Neither a destination nor a place easily found by travelers in the 1930s, the site struggled to attract visitors and stay open. Then Garnet Carter, who developed the first miniature golf course, conceived of using

People who regularly travel Frazier Hollow Road in Dowelltown (DeKalb County) may not know that this bull has been staring from the side of the barn since 1951, when builder W. O. Womack first had the sign painted. Jimmy Womack owns the Dry Creek community landmark today. *Submitted and photographed by Tammy Crook.*

This Sullivan County barn provides a practical and eye-catching display for feed signs. *Submitted and photographed by Anthony Perkins.*

Ruby Falls is a Chattanooga attraction that first opened in the 1930s. Known as the "Jewel of Lookout Mountain," Ruby Falls was named after the wife of spelunker and site developer Leo Lambert. The waterfall is 145 feet high and located 1,120 feet inside the mountain. Only a few of the original "Ruby Falls" signs remain; this one shares billing with a local business on Pete Denning's barn in Robertson County. *Submitted and photographed by Tommy O. Jackson.*

Ernie Smith's barn on River Road in Ashland City (Cheatham County) is a popular subject for photographers. What title could be more appropriate for this wintry scene than "Ice Cold Double Cola?" *Submitted and photographed by John Werne.*

The Hotel Windsor, located in downtown Johnson City, was within walking distance of three passenger railroad stations during its heyday from about 1912 to 1925. A few signs, including this one on a Unicoi County barn, survive the landmark building that was razed in 1971. *Submitted and photographed by John Melton.*

barns as billboards to point the way to visitors from across the country to Rock City. His idea was and still is a marketing success story. Clark Byers, who painted most of signs, recalled that farmers were sometimes paid between $3 and $5 per year. These famous billboards are found in Tennessee and other states and are the subject of a book written and photographed by David B. Jenkins. "See Rock City" barns are among the best-known examples of roadside advertising architecture in the country.

Tennessee Century Farms

Across Tennessee, the yellow metal Century Farm sign is displayed by a select group of farmers who can prove that their land has been owned and farmed continuously by their family for at least 100 years. These farm families represent the state's agricultural history and maintain some of its most historic buildings and rural landscapes. The Century Farms program was begun by the Tennessee Department of Agriculture (TDA) as a way to celebrate the nation's bicentennial in 1976. About 600 farms were in the initial group that received the first yellow signs. In 1984, the Center for Historic Preservation at Middle Tennessee State University was asked by TDA to administer this important collection of stories, photographs, and family and state history, and the Center also promotes the conservation and continued agricultural use of these significant places. The first collaboration between the Center and TDA was *Tennessee Agriculture: A Century Farms Perspective* by Carroll Van West that was published in 1986. It featured stories and photographs of almost 700 farms. Over the past 20 years, the program has

This "See Rock City" barn is located in Humphrey's County, one of many such famously painted barns across the state. *Submitted and photographed by Jeania Wright.*

Listed on the National Register of Historic Places, the Shelley Farm, founded 1839, is the only certified Century Farm in Hamilton County. *Courtesy of the MTSU Center for Historic Preservation.*

grown to include more than 1,200 farms, many of which are more than 200 years old. TDA continues to award signs to Century Farm families. A Web site at http://histpres.mtsu.edu offers information on becoming a Century Farm as well as brief histories and photographs, when available, of all certified Tennessee Century Farms. For further information on this program, contact the Center for Historic Preservation, Box 80, MTSU, Murfreesboro, TN 31732, or call 615-898-2947. Several Century Farm barns are featured throughout this publication.

Appalachian Quilt Trail

The Appalachian Quilt Trail (AQT) in upper East Tennessee features painted quilt squares that are attached to barns marking several routes throughout participating counties. Visitors are invited to drive, walk, or cycle through the countryside visiting towns, communities, and attractions along the trail. As a heritage tourism and educational effort, the Appalachian Quilt Trail is a successful model for other regions of the state. For more information, visit the Web site at http://www.vacationaqt.com/ or contact AQT at 7995 Rutledge Pike, P.O. Box 379, Rutledge, TN 37861, or 1-888-775-4AQT (toll free) or 1-865-828-4586.

Around 1928, this circa 1910 barn was moved to its current location on State Route 411 near Madisonville in Monroe County. Owned by Jon, Cyndi, and Myrna Sharpe, the photo features four generations of the descendents of Daniel Wesley Cardin on the family farm he founded. Lena Watson, standing center, is the granddaughter of Cardin. Her daughter, kneeling to her left, is Myrna Sharpe. *Submitted by Jon Sharpe. Photographed by Angel Jennings.*

Two advertisements that recall the early years of automobile travel are featured on State Route 76 near White House (Robertson County). The photographer notes that a full set of Burma Shave signs, a clever roadside advertising scheme that was an American tradition from the mid-1920s until the early 1960s, and an early Gulf service station can be seen from the lane leading to this farm. *Submitted and photographed by E. E. Price.*

Yellow pine logs, some 60 feet long, were cut from the property and hewn by Ed Colman Franks to build this barn in 1917. It is one of the historic buildings on the Dodd Century Farm (founded 1893) in Hardin County. Franks was the father of current owner Muriel Franks Bradley and grandfather of the photographer. *Submitted by Muriel F. Bradley. Photographed by John F. Bradley.*

McDonald Craig of Perry County stands in front of the barn he built in the 1950s after acquiring the farm that was established by his great-grandparents, Tapp and Amy Craig. On Christmas Day of 1871, they offered $150 and a yoke of oxen as a down payment on this property, one of the first farms owned by freed slaves in the state. Continuing his ancestors' quiet but determined quest for equal rights and opportunities, McDonald Craig transported black students from Perry County to Montgomery High School in Lexington (Henderson County) from 1955 to 1965. The bus that Craig drove for those 10 years is parked in the barn. Craig, a well-known musician, composer, and a member of the Mid-America Country Music Hall of Fame, displays a concert banner, his Century Farm sign, and an oxen yoke he made to symbolize the beginning of his farm that is listed on the National Register of Historic Places. *Submitted by McDonald Craig. Photographed by Larry Lofton.*

This Sullivan County barn on the grounds of the historic Allendale Mansion, owned by the City of Kingsport, displays one of the colorful squares of the AQT. *Submitted and photographed by Michael Murphy.*

A Hawkins County Century Farm. *Submitted by Muriel F. Bradley. Photographed by John F. Bradley.*

The McNews, above, explain that the barn behind them was built in 1954 to replace a 1918 barn. "Garnieta's Star" is the pattern displayed as part of the Appalachian Quilt Trail in Claiborne County. *Submitted by Joe and Garnieta McNew. Photographed by Linda Carter.*

Sammy Lipe has been restoring this early 1900s barn on State Route 70 North near Rogersville (Hawkins County). It is a stop on the AQT. *Submitted and photographed by Sammy Lipe.*

Appropriately, this barn supplies an enduring flag for the community of Independence in Overton County. *Submitted and photographed by Heather Montjoy.*

For nearly 50 years, Max Speiss housed his horses in this barn, which was already on the property when he purchased his Lewis County farm. Mrs. Speiss continues to live on the farm and "gets great pleasure looking out her kitchen window at the barn." *Submitted and photographed by Connie Speiss Hensley.*

If you travel along State Route 96 East near the Rutherford and DeKalb County line, you can hardly fail to notice this hay barn owned by Hugh and Barbara Lester with the American flag front and center. *Submitted and photographed by Lyn Ackley.*

During the years he owned the farm, Jim Cobb had the Tennessee state flag painted on the side of his barn near Spring City (Rhea County). The barn was built by Morgan Cunningham in the 1960s. The current owners are Mr. and Mrs. Rick Latshaw; she is the great-niece of Cunningham. *Submitted and photographed by Suzanne Cobb.*

Flags

In addition to providing backdrops for signs, advertisements, and art, barns are also used by many owners to display either actual or painted versions of the United States and Tennessee flags.

Built in 1962, this Stewart County barn makes a good canvas for a painting of the U.S. flag. *Submitted by Joe and Bernice McClure. Photographed by Angela Elliott.*

The Stars and Stripes hangs proudly on this barn near Dry Fork in Wilson County. *Submitted and photographed by Amy Waggoner.*

An All-American barn sits on the Harpeth Westover Farm in Williamson County. *Submitted and photographed by Christopher Cuomo.*

Loyd Burks shows off his collection of license plates decorating the side of his circa 1960 barn. *Submitted and photographed by Jo Smoot.*

Nature's artwork decorates this barn on a Century Farm in Wilson County. *Submitted and photographed by Glenda Loveless.*

It's the Fourth of July at Paw Paw's in Wilson County. Larry Finch poses with grandson Matt and a feline friend. *Submitted and photographed by Gail Porter.*

Then and Now

The following pairs of photographs document the appearance of outbuildings over time. Some have been maintained or adapted for other purposes, and others are either undergoing repairs or have been reclaimed through careful restoration. These barns were important to their owners then and continue to be useful now.

From the 1930s until 2006, this barn was a part of the Blanton Stock Farm in Bedford County. The restoration was completed in 1986 by a Mennonite crew who found it to be "only 1.25 inches out of square," says Charles McDonald, a former owner. The barn is currently owned by C & N Properties. *Submitted by and current photograph by Colleen Adcock. Vintage photograph supplied by Charles McDonald.*

For nearly 100 years, this Tipton County barn has stood in the Charleston-Gift community east of Covington. Built in 1910 by the Hickman brothers, two of whom are seen in the photograph, the barn is now used "mostly for storage and a place cats and owls love to hang out," report owners Tammy and Keith Davis. *Submitted by and current photograph by Tammy Davis.*

The Hearn family of Wilson County built this barn around 1896 to "house miniature ponies," current owner Howard Gillette says. He bought the property in the 1970s to keep it from being demolished. The barn has been adapted to serve as "Howard's Shop." Daughter Tina is the young lady in the loft in the older photo. *Submitted by Howard Gillette. Photographed by Julia Gillette Chaffin.*

Phil Morris Baggett has restored seven barns on his Montgomery County farm since 1998. The horse barn was built in 1944 by Robert Rittenberry. *Submitted and photographed by Phil Morris Baggett.*

It took two years and a lot of help from family and friends for owners Kelley and Michael Hendrix to restore their barn, built in 1965 by W. D. Elliott, in Cowan (Franklin County). *Submitted by Kelley Hendrix. Photographed by Michael Hendrix.*

In an image from 1905, Josie Combs, Anita Godwin Taylor's grandmother, stands by the horse while her sister, Willie, is mounted to ride sidesaddle. Built in the 1880s by J. Hal Lambert, the corncribs of this Dyer County barn are still intact. The old photograph collection is invaluable to restoration efforts that are underway by the Godwin family. *Submitted by and current photograph by Anita Godwin Taylor.*

The Prater Century Farm in Knox County, part of a plantation of more than 1,000 acres established in 1801 by Benjamin Prater, raises cattle, milk goats, hay, and horses. The 19th-century smokehouse has been remodeled by owner Elsie Prater to serve as an office and showcase for some of her crafts and farm memorabilia. *Submitted and current photograph by Elsie Prater.*

New Uses

Because of their location, materials, craftsmanship, and spacious design, sometimes barns are converted to serve other purposes. These reincarnations prolong the lives of the buildings and allow them to continue to be a part of the landscape and everyday life. Such creative and functional adaptations are examples of recycling at its best as new uses are made of spaces that were once stalls, lofts, and passageways.

Agritourism

In recent years, agritourism has become a viable option for farmers to diversify, add value, and generate additional income while providing unique experiences to visitors. Seasonal crops like pumpkins and Christmas trees, corn mazes, pick-your-own products, fisheries, wineries, bed-and-breakfast accommodations, farm tours, and farms and ranches that feature exotic farm animals including llamas, emus, and ostriches are some of the mainstays of this widespread initiative. For more information on agritourism and alternative farming enterprises in Tennessee, contact either of these agencies:

Tennessee Department of Agriculture
P. O. Box 40627
Nashville, TN 37204; 615-837-5081
or e-mail marketing.staff@state.tn.us

Center for Profitable Agriculture
P. O. Box 1819
Spring Hill, TN 37174-1819
931-486-2777
cpa.utk.edu

The Girl Scouts of Middle Tennessee are enthusiastic about the "Big Red Barn" that has been converted to serve as the heart of Camp Sycamore Hills in Cheatham County. *Submitted by the Girl Scouts of Middle Tennessee. Photographer unknown.*

Polk County is home to the Webb Brothers Century Farm (1880) on the Hiwassee River. The Webb Brothers Float Service is one of the earliest and most successful agritourism ventures in the state. *Courtesy of the MTSU Center for Historic Preservation. Photograph courtesy Ingrid Buehler, Polk County News.*

A recycled dairy barn in Sevier County provides a display area for antiques. *Submitted and photographed by Tommy O. Jackson.*

BARN AGAIN!

BARN AGAIN! is a national program that provides information to help owners of historic barns rehabilitate them and return them to productive use. Through demonstration projects, case studies, publications, technical assistance, and an awards program, BARN AGAIN! is asking that people reconsider, for a variety of reasons, the widely accepted premise that new is better. The program has shown how historic barns can be adapted for new farming uses ranging from dairy, hog, and cattle operations to machinery or grain storage.

Barn preservation can be a cost-effective alternative to demolishing an old barn and constructing a new building. The program provides advice and information to barn owners across the country. BARN AGAIN! is sponsored by the National Trust for Historic Preservation and *Successful Farming* magazine. Contact BARN AGAIN! at the National Trust for Historic Preservation, 535 16th Street, Suite 750, Denver, CO 80202, or 303-623-1504. E-mail barnagain!@nthp.org or visit the BARN AGAIN! Web site at www.barnagain.org.

In 1996, Judy and Glen Clements of Lafayette (Macon County) realized their days of growing tobacco as a primary cash crop were past. Joining other farmers in the growing wine industry of Tennessee, they became vintners and remodeled their former tobacco barn, which is now the headquarters of the Red Barn Winery. *Submitted and photographed by Ginger Riley.*

Income Tax Credits for Barn Owners

Owners of historic barns that are eligible for or listed on the National Register of Historic Places may apply for a 20-percent income tax credit for expenses associated with certified rehabilitation. Because farm buildings are a part of the income production of the farm, they qualify for this income tax credit if specific guidelines are followed in the rehabilitation. For further information on this program, contact the Tennessee Historical Commission, Clover Bottom Mansion, 2941 Lebanon Road, Nashville, TN 37243-0442, or call 615-532-1550.

Pleasant View Century Farm in Williamson County, owned by the Gentry family, is an early and successful model for agritourism. By the 1970s, the family was providing tours and special events for school groups and offering pick-your-own products. The former 1950 tobacco barn houses a maze that is enjoyed by the many school groups that visit the farm. *Submitted and photographed by Barb Jones.*

Owners Steve and Joyce Cook have spent the last eight years making a home of this 1940 barn near Martin in Weakley County. *Submitted and photographed by Trenia Reagan.*

Conservation Easements

Several agencies and not-for-profit organizations across Tennessee can acquire and hold conservation easements that protect land from development. One is the Land Trust for Tennessee, a nonprofit charitable organization with a mission to "preserve the unique character of Tennessee's natural and historic landscapes and sites for future generations." The Land Trust works with willing landowners to preserve the historic and natural values of their land as well as the existing scenic vistas. The primary tool to accomplish this goal is a conservation easement donated by the landowner to the Land Trust, which then holds it in perpetuity. The landowner continues to own, use, or sell the property, but within the terms of the agreement set forth by the easement. For more information, contact the Land Trust for Tennessee, 209 10th Avenue South, Suite 530, Nashville, TN 37202, or call 615-244-5263. You also may visit its Web site at www.landtrusttn.org. Many landowners throughout Tennessee are electing the conservation easement as a tool to forever preserve their land, or parts of it, from development.

The Donnell Century Farm, founded in 1825 near Jackson (Madison County), is a working farm that also offers educational field trips. From the hay loft of the newly restored 1920s mule barn, Billy and Rose Ann Donnell and their family invite visitors to participate in the Donnell Century Farm Adventure. *Submitted and photographed by Larry Atherton.*

The old granary on Still Hollow Century Farm, founded in 1857, now serves as The Farmer's Wife Gift Shop as the Birdwells of Greene County diversify to include agritourism options. *Submitted and photographed by the Birdwell family.*

The 700 acres of the original Mayfield Century Farm founded in 1833 in McMinn County, home of Mayfield Dairy, has been placed into a conservation easement with the Land Trust for Tennessee. The landscape of pasture lands and hardwood trees will be preserved forever. Plans call for an agritourism effort to begin soon that will focus on ornamental horticulture and education that will be linked to the tour of the nearby dairy facilities. *Submitted and photographed by the Land Trust for Tennessee.*

Almost Gone

"Like anything old you stand now in quiet, reflective dignity. Even the dilapidation reflected in your appearance adds authenticity to what you have, at one time, been. You should remind the viewer that regardless of the more rapid pace of modern life it is, in reality, merely an extension of what has gone before, it is not something new. Maybe you should remind modern viewers that structures like yourself are more permanent than their hurried efforts to make a mark on the world. And this also, that at least one true measure of significance is the influence that lives after us." — From *The Disappearing Landscape* by Bob Womack.

Tennessee's landscape is scattered with barns that seem to have outlived their usefulness and patiently wait for nature or people to dismantle them entirely. As development surrounds them, as their structure is weakened by neglect, wind, and rain, and as weeds and vines cover them, they have a sobering effect on passersby who capture their final years or days in photographs, paintings, and poetry. These places, which now shelter small creatures and store old tools, equipment, rotting bales of hay, or perhaps a lone piece of harness, are poignant reminders of a past that was alive with the sounds, sights, and smells of a busy agricultural lifestyle. Today, whatever the reasons for their imposed solitude and decline, these remnants of once-active farmsteads undeniably still influence us even though their future is uncertain.

The small building on the left is the former Jingo Post Office (Williamson County) that was moved to this site and used for milk storage when the farm supported a dairy herd. Plans call for the building to be moved to another location and restored. The barn, built in 1945 by Howard Elrod, and the silo may "give way to a major housing development planned for this site," writes Randall B. Nash. *Submitted and photographed by Randall B. Nash.*

As this Lake County barn is lost to view more each year, the photographer writes that it is as if "the vegetation is slowly swallowing her like a ship sinking beneath the waves." *Submitted and photographed by John E. Dial.*

"Barns, old and grey, eroded by time
Haunting shadows of the past
Faithful sentinels well past their prime."

Quote submitted by Iva Lee Wolfe.

A grain drill registered in 1896 was found in this barn, built by Nancy Fessenden's brother in 1903 in Madisonville (Monroe County). *Submitted and photographed by Nancy Fessenden.*

As residential development crowds it, the fate of this barn in Morristown (Hamblen County) is in question. *Submitted and photographed by Rick Bunge.*

A dazzling pattern of light filters through the exposed rafters that were raised by Glenda Cantrell's great-grandfather, Joe Argo, in Grundy County in the 1890s. The family still has his woodworking tools. *Submitted and photographed by Glenda Cantrell.*

Nature has nearly claimed this Madison County barn that can be seen from the bypass north of Jackson. Jay Moore, son of the photographer, was inspired to write a poem titled "The Abandonment," which concludes with these lines: "It's slowly destroyed by the heat and the cold, / But nobody cares, because the barn is so old." *Submitted and photographed by Mary N. Moore.*

Just a Memory

Nature, neglect, age, wear over time, decisions to utilize the land and materials in other ways, and sometimes, regrettably, arson all take aged outbuildings from the landscape. Barns that no longer exist often remain in the thoughts of people who recall their unmistakable silhouette on a hillside or their dominating presence in the farmyard. We remember families and workers who built and used them over the decades and the animals that were sheltered within their stalls and fed from their lofts and cribs. These buildings become so familiar to us that their absence from the space they so long occupied both surprises and saddens us again and again. The following photographs represent the great barns of Tennessee that no longer exist except in pictures and memory.

The barn built by Bill Allen and owned by John Allen in Carter County was claimed by a storm in 2006. *Submitted and photographed by Randy C. Whaley.*

As barns are stripped to their skeletal framework, they reveal the skill and materials involved in their construction. This Rutherford County barn on New Salem Road and Warrior Drive near Interstate 24 caught the interest and eye of many passersby when it was being demolished in 2007. *Submitted and photographed by Janice McCartney.*

Built-in mangers gave this Williamson County barn its distinctive shape. Like the storm that was fading when this photograph was taken, this 1950 barn vanished from Wilson Pike in 2003. *Submitted and photographed by Vernon Taylor.*

Round barns are a readily recognizable barn type across the country, though few were built in Tennessee. Several, however, were constructed by Swiss immigrants in the southeastern part of the state. Round barns, some with central silos, were most often built when horses and humans were the primary sources of power. With the emergence of tractors, haybalers, and other mechanized equipment, round barns became awkward to work from, and few were built after the 1920s. This example was part of the Elm Grove Farm, owned by Robert Wilkinson, in Franklin County. The photograph was included in the "Belvidere Community Improvement Scrapbook" from the 1950s, recalls Fred V. Warmbrod, who lives in the Belvidere community and owns the East View Century Farm. Mr. Warmbrod says that no round barns are known to exist in the area today. *Submitted by Fred D. Warmbrod. Photographer unknown.*

Sharing her memories of "Grandaddy's Barn" in Gibson County, Donna Roberts recalls this barn as a place where "hundreds of cows were milked" and where "grandkids loved to play in the hayloft." Owned by Raymond Pugh, the barn was built in 1890 by Bob Ramsey and neighbors. A tornado ended its 116 years of service in 2006. *Submitted and photographed by Donna Roberts.*

The photographer explains that this beautiful horse barn was located at the Kennesaw Farm about four miles southwest of Gallatin on Highway 31E (Nashville Pike). The barn was to be removed for a big construction project and was eventually burned for the documentary movie *80 Acres of Hell* about the Civil War Camp Douglass. *Submitted and photographed by Allen Haynes.*

Haze Alexander built this Claiborne County barn in the 1930s. A part of Frank Heck's farm in historic Powell Valley, it was the victim of a spring 2007 storm. *Submitted and photographed by Jana Stover.*

A landmark for decades on the Arno-Allisona Road in Williamson County, the photographer recalls that this barn had a "small stream flowing through it." A tornado helped it on its way to demolition. *Submitted and photographed by W. C. Scarlett Jr.*

The photographer, who lives in Portland (Sumner County), remembers that, "This barn stood at the top of a hill, looking out over the countryside. I always looked for it when I drove into town. It was recently torn down." *Submitted and photographed by Carolyn Whittenburg.*

Highway 64 now covers the former site of this McNairy County barn. Whit Ash and his neighbors built the barn some 70 years ago. *Submitted and photographed by Larry Atherton.*

Bibliography

Books and Articles

Apps, Jerry. *Barns of Wisconsin*. Madison: Tamarack Press, 1977.

Arthur, Eric, and Dudley Witney. *The Barn: A Vanishing Landmark in North America*. New York: Arrowood Press, 1988.

Betts, Edwin Morris, ed. *Thomas Jefferson's Farm Book: With Commentary and Relevant Extracts from Other Writings*. Richmond: University Press of Virginia, 1987.

Ensminger, Robert F. *The Pennsylvania Barn: Its Origin, Evolution, and Distribution in North America*. 2d ed. Creating the North American Landscape Series. Baltimore: Johns Hopkins University Press, 2003.

Glassie, Henry. "The Old Barns of Appalachia." *Mountain Life and Work* 41 (Summer 1965): 21-30.

———. "The Double Crib Barn in South Central Pennsylvania." *Pioneer America* 1 (January 1969): 9-16; 1 (July 1969): 40-45; 2 (January 1970): 47-52; 2 (July 1970): 23-34.

Afternoon sunlight filters across the name carved into a feed room door of Toni and Robert Conover's barn near Bradford in Gibson County. *Submitted and photographed by Robin Conover.*

———. "The Variation of Concepts Within Tradition: Barn Building in Ostego County, New York." *Geoscience and Man* 5 (June 1974): 177-235.

Hardeman, Nicholas P. *Shucks, Shocks, and Hominy Blocks: Corn as a Way of Life in Pioneer America*. Baton Rouge: Louisiana State University Press, 1981.

Jacques, D. H. "Barns, and Other Out-Buildings." Chap. in *The House: A Manual of Rural Architecture: Or, How to Build Country Houses and Out-Buildings*. New York: Geo. E. Woodward, 1869.

Jenkins, David B. *Rock City Barns: A Passing Era*. Chattanooga: Free Spirit Press, 1996.

Jordan-Bychkov, Terry G. "Transverse-Crib Barns, the Upland South, and Pennsylvania Extended." *Material Culture* 30 (No. 2 1998): 5-31.

Kniffen, Fred. "Folk Housing: Key to Diffusion." *Annals of the Association of American Geographers* 55 (No. 4 1965): 549-577.

Larkin, David. *Farm: The Vernacular Tradition of Working Buildings*. New York: Monacelli Press, 1995.

Leffingwell, Randy. *The American Barn*. St. Paul: Motorbooks International, 2003.

Marshall, Howard Wight. *Barns of Missouri: Storehouses of History*. Virginia Beach: Donning Company, 2003.

McKenzie, Robert Tracy. *From Old South to New South in the Volunteer State: The Economy and Society of Rural Tennessee, 1850–1880*. Ann Arbor: University Microfilms International, 1991.

———. *One South or Many?: Plantation Belt and Upcountry in Civil War-Era Tennessee*. New York: Cambridge University Press, 1994.

Moffett, Marian, and Lawrence Wodehouse. *East Tennessee Cantilever Barns*. Knoxville: University of Tennessee Press, 1993.

Montell, William Lynwood, and Michael Lynn Morse. *Kentucky Folk Architecture*. Lexington: University Press of Kentucky, 1976.

Morgan, John, and Ashby Lynch Jr. "The Log Barns of Blount County, Tennessee." *Tennessee Anthropologist* 9 (No. 2 1984): 85-103.

Morrison, Alice Reed. "Ethnicity and Acculturation: German Immigrant Homes and Barns of Southern Indiana." *Material Culture* 34 (No. 1 2002): 1-39.

Noble, Allen G. *Wood, Brick, and Stone: The North American Settlement Landscape. Vol. 2, Barns and Farm Structures*. Amherst: University of Massachusetts Press, 1984.

———, and Richard K. Cleek. *The Old Barn Book: A Field Guide to North American Barns and Other Farm Structures*. New Brunswick: Rutgers University Press, 1995.

Pillsbury, Richard, and John Florin. *Atlas of American Agriculture: The American Cornucopia*. New York: Simon & Schuster Macmillan, 1996.

Price, H. Wayne, and Keith A. Sculle. "Gothic Barns in Brown County, Illinois: A Survey and Interpretation." *Material Culture* 30 (No. 3 1998): 1-41.

Rawson, Richard. *Old Barn Plans*. New York: Bonanza Books, 1982.

Riedl, Norman F., Donald B. Ball, and Anthony P. Cavender. *A Survey of Traditional Architecture and Related Material Folk Culture Patterns in the Normandy Reservoir, Coffee County, Tennessee*. Report of Investigations No. 17, Department of Anthropology, University of Tennessee, Knoxville. Knoxville: Tennessee Valley Authority, 1976.

Roe, Keith E. *Corncribs: In History, Folklife, & Architecture*. Ames: Iowa State University Press, 1988.

Sculle, Keith A., and H. Wayne Price. "The Traditional Barns of Hardin County, Illinois: A Survey and Interpretation." *Material Culture* 25 (No. 1 1993): 1-27.

Sharp, Leslie N. "Down South to Dixie: The Development of the Dixie Highway from Nashville to Chattanooga, 1915–1940." Master's thesis, Middle Tennessee State University, 1993.

Simmonds, William G. *Advertising Barns: Vanishing American Landmarks*. St. Paul: Motorbooks International, 2004.

Sloane, Eric. *An Age of Barns: An Illustrated Review of Classic Barn Styles and Construction*. Stillwater, Minn.: Voyageur Press, 2001.

———. *American Barns and Covered Bridges*. New York: Funk and Wagnalls, 1954.

Vince, John. *Old Farms: An Illustrated Guide*. New York: Bramhall House, 1986.

Visser, Thomas Durant. *Field Guide to New England Barns and Farm Buildings*. Hanover, N.H.: University Press of New England, 1997.

Vlach, John Michael. *Barns*. Library of Congress Visual Sourcebooks in Architecture, Design, and Engineering Series. W. W. Norton & Company, 2003.

West, Carroll Van. *Tennessee Agriculture: A Century Farms Perspective*. Nashville: Tennessee Department of Agriculture, 1986.

———, and Connie L. Lester, Margaret Duncan Binnicker, and Anne-Leslie Owens, eds. *The Tennessee Encyclopedia of History and Culture*. Nashville: Tennessee Historical Society, 1998. Also available online at http://tennesseeencyclopedia.net/

A Sumner County barn near Goodlettsville. *Submitted and photographed by Robin Conover.*

Winters, Donald L. *Tennessee Farming, Tennessee Farmers: Antebellum Agriculture in the Upper South*. Knoxville: University Press, 1994.

Womack, Bob. *The Echo of Hoofbeats*. 2d ed. Shelbyville, Tn.: Dabora, Inc., 1984.

———. *The Disappearing Landscape*. Murfreesboro, Tn.: Bob Womack, n.d.

Web Sites

American Shire Horse Association
http://www.shirehorse.org/

Dog Breed Info Center
http://www.dogbreedinfo.com/mountainfeist.htm

Gentle Giant Ranch
http://www.angelfire.com/ca/GentleGiantShires/

University of Tennessee Agricultural Extension Experiment Station Highland Rim Research and Education Center
http://highlandrim.tennessee.edu/

University of Tennessee Extension
http://www.utextension.utk.edu/

University of Tennessee Institute of Agriculture
http://www.agriculture.utk.edu/news/releases/0711-CenterforTobacco.htm

USDA, National Agricultural Statistics Service, Tennessee Field Office
http://www.nass.usda.gov/tn

Index

Nailed to a barn door in the Cable Mill Historic Area of Cades Cove in the Great Smoky Mountains National Park, this masterfully forged horseshoe hinge attests to a blacksmith's handiwork. *Submitted and photographed by Robin Conover.*

Countless baskets were made using this Springfield barn as a backboard. *Submitted and photographed by Robin Conover.*

Used as a cattle barn today, this Sumner County barn is located on Patton Branch Road. *Submitted and photographed by Robin Conover.*

About the Cover Photographs

Front cover

Built-in mangers gave this Williamson County barn its distinctive shape. Like the storm that was fading when this photograph was taken, this 1950 barn vanished from Wilson Pike in 2003. *Submitted and photographed by Vernon Taylor.*

Left inside front cover from top left to bottom right

A stock barn in Jackson County. *Submitted and photographed by Pete Konrad.*

Shelby County is the home of a tidy gambrel-roofed barn with flanking side sheds. *Submitted and photographed by Charles Welch.*

A tobacco barn in Cheatham County. *Submitted and photographed by Carolyn Whittenburg.*

Cal Turner Jr. began a Christmas tradition in 1996 when he commissioned an artist to create a nativity scene. Painted on canvas, the mural is rolled onto the barn roof each year. The Turner farm is located on Franklin Road near Brentwood in Williamson County. *Submitted by Barbara Kinser.*

Arvil Holt built this Wayne County barn in 1953. Four generations later, it is still owned by the family. *Submitted by Karla Roman. Photographed by Jeff Roman.*

This Jefferson County barn, locally known as Papaw Burchfiel's barn, is located in Talbott. *Submitted by Lura Burchfiel Hinchey. Photographed by Sharon Knechtel.*

Moffitt Farm's working barn in Warren County near McMinnville. *Submitted by Bill Moffitt. Photographed by Leah Lindsey Moffitt.*

Afternoon sunlight on a Sumner County barn. *Submitted and photographed by George Hornal.*

Sugarfoot Farm in Maury County. *Submitted and photographed by Lee Keck.*

A Hamblen County barn near Bull's Gap. *Submitted and photographed by Lynn Huffstetler.*

A Dickson County barn. *Submitted and photographed by Carlene Bennett.*

A Lawrence County horse barn. *Submitted and photographed by Kathy Boice.*

Loudon County barn. *Submitted and photographed by Krista Freeman.*

A central passage barn in Williamson County. *Submitted and photographed by Thomas Thomas.*

This nicely detailed red hay barn stands in Putnam County. *Submitted and photographed by Allen Haynes.*

As storm clouds pass, a rainbow appears over this DeKalb County barn. *Submitted and photographed by Lonnie Cantrell.*

Right inside cover from top left to bottom right

This well-maintained barn is located in Humphreys County. *Submitted and photographed by Margaret Watson.*

Locals say the springhouse below the main barn on this Cannon County cattle farm has never run dry. *Submitted and photographed by Anne Goetze.*

Built in a large pasture, this Meigs County barn has provided shelter and storage for decades. *Courtesy of the MTSU Center for Historic Preservation.*

Union County barn. *Submitted and photographed by Darrell Beason.*

Fall color brightens the landscape around this Hamblen County barn. *Submitted and photographed by Sandra Murray.*

This Cocke County barn is located near Newport. *Submitted and photographed by Joy Reidy.*

The view of a working farm from the hallway of a Cumberland County barn built by Bill Garrett. *Submitted and photographed by Ken Johns.*

This Morgan County barn was built by the Melhorn family more than 50 years ago. *Submitted and photographed by Tyler Landrum.*

Bordered by a rock fence, this Trousdale County barn was built by Hubert Ward in 1910. *Submitted by Edgar Waller. Photographed by Frances Waller.*

Horses graze in the pasture in front of this Monroe County barn located near Madisonville. *Submitted and photographed by Pamela Cox.*

Built in the 1933, this Knox County barn has remained in the same family for three generations. *Submitted and photographed by Scott and Charlotte Shackelford.*

Built in 1949 for Raford Sewell and Bobbie Faye Nobles of Jacks Creek in Chester County, this is the last of the Nobles' family barns still standing. *Submitted and photographed by Patsy Nobles Jones.*

Working barns in Putnam County. *Submitted and photographed by Janice McCartney.*

James Sowell built this Maury County barn located in Santa Fe. *Submitted and photographed by Kathy Chunn.*

Neighbors helped Leonard Parham build this Gibson County barn more than 75 years ago in the Locust Grove community. It replaced the original barn, which had been struck by lighting and burned to the ground. *Submitted by Monette Parham. Photographed by Linda Parham.*

The Swint family of Morgan County built this barn more than 100 years ago. *Submitted and photographed by Tyler Landrum.*

Left inside back cover from top left to bottom right

Built in the Sequatchie Valley more than 50 years ago, this working barn is located in Bledsoe County. *Submitted and photographed by Howard Blair.*

A Benton County barn stands in the late afternoon sunlight. Built by A. T. and Fred Cuff some 70 years ago, the barn has sheltered horses, cattle, and pigs. *Submitted and photographed by Beverly Hicks.*

A hard-working little farmer, Jonas Vickers Hannah is the great-grandson of Marvin Vickers, who built this Coffee County barn in 1942. *Submitted by Reba Vickers Cochran. Photographed by Larry Vickers.*

The fertile soil of Claiborne County's Powell Valley remains a primary attraction to farmers after more than 200 years. Surrounded by a network of fencing, this multipurpose barn is owned by Jeffrey Ellison. A ramp built into this central hall-style barn allows a team to pull a wagon full of loose hay into the barn to store it. *Submitted by the Tennessee Farmers Cooperative.*

A double rainbow frames this Maury County barn. *Submitted and photographed by Richard Mullen.*

A Maury County barn with grain silo. *Submitted and photographed by Mae Wilhelm.*

A Century Farm in Grainger County that is also on the Appalachian Quilt Trail. Clinch Mountain can be seen in the distance. *Submitted and photographed by R. F. Ellis.*

This Chester County barn was built from the debris of three homes destroyed in a tragic 1952 tornado. The 8,500-square-foot structure was built 56 years ago to honor those killed in the storm. This structure, unfortunately, burned in January 2008. It has since been rebuilt. *Submitted and photographed by Jerry Thornthwaite.*

This Lincoln County farm was once owned by Collier Barnes, who farmed and raised cattle here for many years. The barn stands near Petersburg and is more than 80 years old. *Submitted and photographed by Kristi Caves-Bartlett.*

A Hickman County barn on a quiet country road near Primm Springs. *Submitted and photographed by Renee Lowery.*

Turkey and deer can often be seen near this Williamson County barn. *Submitted and photographed by Dinah Freeman.*

Built in 1937 by the V. V. Snead family, this Obion County barn was used to store hay and corn for the cattle farm. Sammy Snead fully restored his grandfather's barn in 1998. *Submitted by Sammy Snead. Photographed by Beth McDavid.*

The painted roof of this Unicoi County barn advertised Wiseman's Clothing and Shoes in nearby Johnson City. *Submitted and photographed by John Melton.*

An inside view of the tiers used to hang tobacco inside of a Claiborne County barn. *Submitted and photographed by Pat Williams.*

A brick barn built in Davidson County near Joelton. *Submitted and photographed by Barbara Pupak.*

Red-winged blackbirds land in front of this Sumner County barn. *Submitted and photographed by Carolyn Whittenburg.*

Right inside back cover
from top left to bottom right

L. L. Tidwell owned this Dickson County farm until his death. Through the decades, the barn has housed cows, mules, and horses. *Submitted and photographed by Shae Beggrow.*

A Johnson County farm in the valley. *Submitted and photographed by John Melton.*

Frank Swallows, his son, Ray, and Ray's father-in-law, Chester Neal, purchased several homes in the Willow Grove community that were to be flooded by the newly constructed Dale Hollow Dam. The men dismantled the homes and used the lumber to build the Swallows' barn on Ivy Hill Farm in Overton County in 1942. *Submitted by Paula Swallows Stover. Photographed by Lisa Elliott Powell.*

William H. Carpenter purchased the farm where this Smith County barn is located in 1870. Carpenter raised and sold cows, sheep, donkeys, turkeys, and chickens for years at his New Middleton Stock Farm. *Submitted by Elsie Minchey. Photographed by Patsy Glunt.*

O. C. Joines and Bert Harrison built this barn in Vervilla about 1938. The Ben Lomand Mountains can be seen in the background of the Warren County barn. Today it is owned by Winford and Eva Sue Bouldin. *Submitted and photographed by Vickie Bouldin.*

Albert Jones built this Robertson County barn before 1900. The barn's stone foundation was quarried from the nearby south fork of the Red River. *Submitted and photographed by Newell Hauser.*

A Grainger County barn built as a freespan-hayloft barn. *Submitted and photographed by Frances Gray.*

This barn stands in Carter County near Elizabethton. *Submitted and photographed by Edward Smelser.*

Arthur D. Barnes' granddaughter, Barbara Wright, discovered that her grandfather built this large Lincoln County barn only after she took this photograph. Built in 1944, the barn remained in the Barnes family for six generations. It is now owned by Tim and Dr. Theresa Morrison. *Submitted and photographed by Barbara Wright.*

Built more than 50 years ago, this dark-fired tobacco barn is located near Barren Plains in Robertson County. It is owned by Corbin Dairy. *Submitted and photographed by Amy Hooberry.*

A Coffee County barn advertising Meramec. *Submitted and photographed by Angel Pardue.*

Locally known as Bob Taylor Parson's barn, this Bedford County barn is currently owned by Randall Crowell. *Submitted and photographed by Colleen Adcock.*

A stone fence borders this Williamson County barn. *Submitted and photographed by Robert McCurley.*

This Grundy County barn near Beersheba Springs was built by J. B. Hill and his son, H. G. Hill, about 1959. It became the center of a large cattle operation. Today the barn is owned and has been fully restored by the Howell E. Adams Jr. family. *Submitted by Howell E. Adams Jr. Photographed by Madeline R. Adams.*

Built and still owned by Howard E. Courtney, this Hamblen County barn dates to about 1970. *Submitted and photographed by Sandra Murray.*

Drury Wallace built this barn more than 150 years ago in Johnson County. Local lore says that Wallace slept in the log structure to prevent his corn from being stolen. *Submitted and photographed by Carolyn Reece.*

Back cover

A hard-working little farmer, Jonas Vickers Hannah is the great-grandson of Marvin Vickers, who built this Coffee County barn in 1942. *Submitted by Reba Vickers Cochran. Photographed by Larry Vickers.*

About the Authors

Caneta Skelley Hankins, a native Tennessean, is the assistant director of the Center for Historic Preservation at Middle Tennessee State University in Murfreesboro. She is a graduate of Martin Methodist College and holds degrees from MTSU. Since 1984, her work with the Center has focused on architectural and cultural preservation of rural areas, research, heritage education for the K-12 grades, and publications. As director of the Tennessee Century Farms program, Hankins has expanded the statewide effort that now documents and recognizes more than 1,200 farms that have been in the same family for at least 100 years. The seventh generation of her family to live in Williamson County, Hankins has been familiar with the general livestock barn (ca. 1930s) on the Skelley farm in Triune all of her life. She is grateful to cousins Mary and Tom Fox for keeping the farm in the family for another generation. Hankins dedicates her part in this publication to the memory of her grandfathers, James T. Grimes, a builder of barns, and Henry Potts Skelley — good farmers and good men.

Photograph by Veita Jo Hampton

Michael Thomas Gavin is the preservation specialist for the Tennessee Civil War National Heritage Area, a statewide program administered by the Center for Historic Preservation at Middle Tennessee State University. Gavin received a B.A. degree in English from Rutgers University and an M.A. degree in History from MTSU. For 25 years, he was president of a construction company that specialized in preserving and rehabilitating historic buildings. As a carpenter and builder of barns as well as a student of rural architecture, he has spent much of his life identifying and restoring traditional houses and farm buildings throughout the state. Gavin's current research areas include African-American history, the charcoal iron industry, and vernacular architecture, particularly log buildings. Some of his recent publications include "From Bands of Iron to Promise Land: The African-American Contribution to Middle Tennessee's Antebellum Iron Industry," *Tennessee Historical Quarterly* 64 (2005), and "Building with Wood, Brick, and Stone: Vernacular Architecture in Tennessee, 1770-1900," in *A History of Tennessee Arts: Creating Traditions, Expanding Horizons* (Knoxville: University of Tennessee Press, 2004). Gavin would like to dedicate his part in this publication to his long-time friend, business partner, and fellow barn-builder Steven Rohrbach, one of nature's noblemen and a true craftsman.

Photograph by Linda Lee Gavin

WISEMAN'S
CLOTHING & SHOES
THIS EXIT